Contract Farming

Editor
Dr. Premjit Sharma

2023
Gene-Tech Books

ISBN 978-81-89729-61-5

Published by **: GENE-TECH BOOKS**
4762-63/23, Ansari Road, Darya Ganj,
NEW DELHI - 110 002
Phone: 43003222
e-mail: genetechbooks@yahoo.co.in

PRINTED IN INDIA

Preface

Contract farming refers to varied formal and informal agreements between producers and processors or buyers. It may include loose buying arrangements, simple purchase agreements and supervised production with input provision, with tied loans and risk coverage. In this system for the production and supply of agricultural/horticultural produce are under forward contracts between producers/suppliers and buyers. The essence of such an arrangement is the commitment of the producer/ seller to provide an agricultural commodity of a certain type, at a time and a price, and in the quantity required by a known and committed buyer. Contract farming usually involves the following basic elements-pre-agreed price, quality, quantity or acreage (minimum/maximum) and time +Contracts can range from oral deals to formal, registered written contracts.

Rapidly expanding demand for high-value food commodities gives farmers incentives to diversify their production portfolio towards commodities that have a strong potential for higher returns. High-value agriculture is best suited to smallholder production systems, as the production of these agricultural commodities requires more labour resources and smallholders have plenty of family labour. It is said that companies prefer to buy from a large number of small farmers because there is less default (even if one or two default, there is assured supply of raw materials), but in reality companies prefer to contract with large farmers in order to reduce the transacting costs (as collection of crops is easier and less costlier from a few large farmers, than with numerous small farmers .As some authors point out, there is evidence of both inclusion of small farmers in contract farming ventures (in resource providing contracts particularly) and exclusion of small farmers (when there are scale constraints). These studies show the necessity for the inclusion of small farmers in the contracts and the contestation whether in reality they are included in all contracting ventures.

Contract farming is prevalent in various parts of India for commercial crops like sugarcane, cotton, tea and coffee. Recently, there have been several private sector initiatives for direct procurement from farmers by companies.. One of the prime advantages for the farmer is that the company will take all the produce grown within specified quality and quantity parameters .. They studied three products that are being contracted from small farmers in North India namely broilers, milk and vegetables, and found out that contract farmers had more reliable market access than non-contract farmers and their income increased manifold in the case of vegetables and milk. In the case of broilers, income was fluctuating. They also argued that accessibility to the markets was one of the criteria that affected participation of small farmers in these contracts.

The present book comprises of 14 chapters. These chapters include introduction to contract farming, specifications of contract farming, managing to contract farming, monitoring performance, impact of contract farming on society, contract farming in developing countries, contract farming and poverty reduction, equity and efficiency in contract farming, distribution in contract farming, coomercialisation of contract farming, consequences of contract farming, contract farming in agri-food system development, advantages and problems of contract farming and successful contract farming.This book introduces the concepts of contract farming with particular reference to India and provides different strategies adopted by various firms and farmers to reap the benefits of contract farming.

In spite of best efforts, it is possible that some errors may have occurred into the writing and editing of the book. Further queries, suggestions and criticisms for improvement of the book are always welcome and shall be thankfully acknowledged.

Premjit Sharma

Contents

1

Introduction to Contract Farming

Traditionally, development in the agricultural sector put emphasis on increasing productivity using external inputs with insufficient attention other aspects, in particular to market linkages. This strategy has resulted in mixed poverty outcomes. With globalisation, market liberalisation, and the development of rural infrastructure, new market opportunities for high-value crops and livestock production are opening up. However, for the rural poor to take advantage of new market opportunities, backward and forward market linkages must be put in place.

These linkages include provision of information on what to grow, rural credit, farming inputs, agricultural extension advice, and help in product accreditation. Putting in place the necessary agri-services for a massive number of small farms and un-organised farmers will require considerable resources from governments. Successes in the provision of public sector agri-services are rare and failures have been numerous. In recent years, a strategy involving private sector has been looked upon as an alternative.

Under contract farming, the purchaser (agri-business firm or trader) provides farmers with inputs, credit, technical advice and market services. In return, farmers produce a certain quantity and quality of crop or livestock, and sell them exclusively to the purchaser. Such arrangements allow farmers to have access to an array of agricultural services, which they would otherwise not have access to.

The emergence of contract farming as an institution for facilitating market exchange is not a recent phenomenon. For decades, contract farming has been used as a supply chain governance strategy in response to market and institutional failures that characterise the agricultural sector in different stages of development. While contract farming itself has been around for a long time, its importance as a tool for transforming subsistence to commercial farmers and thus contributing to poverty reduction has only been reviewed in recent years.

Internationally, in response to changes in consumer preference within developed countries, multinational food corporations are engaging in contract farming in developing countries, mainly to ensure year-round supply of particular product ranges to specific markets and to take advantage of lower production costs. Contract farming would provide the rural poor in these areas with access to a vastly growing export market and hence opportunities to improve their income.

Benefits of Contract Farming

The contract farming identifies several major areas where contract farming can provide benefits. From the point of view of farmers, contract farming can provide access to markets, credit, technology and inputs that they would otherwise be excluded from. Moreover, contract farming can lead to improvements in income while reducing some of the risks they face from production and price fluctuations. From the point of view of purchasers, contract farming provides greater control over volume and quality consistency; to a certain extent, it can also lower certain transaction and production costs that purchasers face.

Table 1. Potential benefits of contract farming

Parameter	Potential benefits
Farmers	
Access to markets	Contract farming arrangements serve to link farmers to distant markets where the demand and price of crops are more favorable.
Access to credit	Purchasers extend credit to farmers either in cash or in kind by providing inputs such as seeds. In cases where purchasers do not extend loans to farmers, banks may accept the contracts as collateral.
Access to technology/ skills development	Contract farming arrangements may facilitate the introduction of new production techniques, and further measures that upgrade agricultural commodities. These include training and assistance in crop production, soil and water management, and bookkeeping of inputs and outputs.
Access to inputs	Purchasers may undertake measures to ensure that contracted producers have timely access to inputs including seeds and fertilisers, in addition to training support and monitoring proper crop husbandry practices.
Increased income	Contract farming can lead to improved income, especially in cases where contract farming is adopted for non-traditional crops that are sold at a premium.

Reduced price risk	In contract farming a predetermined price for the crop is generally established during contract negotiations at the onset of the growing season. This may protect farmers from incurring losses in sales due to downward price fluctuations.
Reduced production risk	Contract farming arrangements facilitate risk sharing from production failures due to uncontrollable circumstances including weather or disease. Purchasers may absorb losses associated with reduced or non-existent throughput for the processing facility. Where production problems are widespread as a result of uncontrollable events, purchasers may defer the repayment of production advances until the following season.
Purchasers	
Control over volume and consistency	Contract farming assures suppliers that the required crops can be produced consistently. Contract farming may result in increased yields and improved quality with regard to certain types of crops.
Improved cost efficiency	Contract farming allows firms to minimise costs by not purchasing land or directly hiring labour.
Contract	farming can help firms minimise supervision costs, usually incurred due to classic principal-agent problems.

Contract farming likewise affords potential benefits to governments. While the development of market linkages for farmers is traditionally viewed as a public sector responsibility, the establishment of the necessary agro-services for a large number of small, unorganised farmers requires a considerable amount of public sector resources. On the other hand, contract farming provides market linkages in ways, which do not burden the public sector.

Contract Farming and Organic Agriculture

In recent years, consumer concerns surrounding food safety have led to an increase in demand for organic products. The global market for organic products has been growing steadily not only in Europe and North America but in Asian countries such as Japan and it is estimated that it will continue to be the fastest growing sector in agriculture. Not surprisingly, organic food production has increased all over the globe, with much of the increase occurring in developing countries where farmers are being attracted by export benefits and substantial price premiums.

This increase in demand has come with a greater insistence on verifiable evidence of food product quality. This in turn has led to more stringent certification requirements and an influx of food traceability systems. Since chemical residues on food are not generally visible and conducting bio-chemical tests are costly and impractical, in order to guarantee the quality of products while minimising transaction costs, certification systems and traceability systems have developed to provide information on products for consumers, notably in developed countries.

For export agents in developing countries, the ability to fulfill the traceability or certification requirements will determine their success in the export market for agricultural products, in particular, high value products such as organic products. Since such products have to meet strict quality requirements that are typically difficult to meet in spot markets, firms are utilising contract farming to gain better control of inputs, achieve more uniform product attributes, and to reduce the cost of measuring quality, grading, and sorting of products.

Due to higher management costs of a food traceability system and the requirement for organic farming to be grown in areas free from chemicals, export firms are likely to engage farmers in marginal areas, where the cost of labour is lower and where use of agrochemicals is minimal. For farmers, contract farming provides access to information and markets. For purchasers, contract farming provides control over inputs throughout all stages of production and processing, making it easier to implement traceability systems. For the government, contract farming facilitates the production of high value agro-products that are consistent with international standards for food safety and traceability.

Beyond issues of establishing the requisite certification standards and traceability systems to promote exports, there are broader benefits from promoting contract farming for organic agriculture for poverty reduction. Over the years, it has become increasingly clear to farmers, non-governmental organisations (NGOs), governments and international development agencies alike that the conventional practice of farming under the banner of the 'Green Revolution' has by-passed the poor in marginal areas while benefited the richer farmers in fertile areas.

There is also increasing evidence that high-external-input agriculture is unsustainable. The unsustainable nature of conventional agriculture is manifesting itself in terms of stagnant or declining yields, increasing ecological degradation, and worsening rural socio-economic conditions. Increasingly, countries have started to look at organic agriculture as a means of reversing these negative effects.

But while the benefits of implementing contract farming may be diverse, and while different stakeholders stand to gain from these benefits, persuading farmers to take part in such an arrangement is still largely a matter of financial incentives. In contract farming, one of the principal motives for smallholders consent is the promise of a steady and increased income incurred from the sale of their crops. To establish the benefits of contract farming providing empirical evidence is crucial.

Types of Contract Farming

Contract farming can be structured in a variety of ways depending on the crop, the objectives and resources of the sponsor and the experience of the farmers. Contracting out production is a commercial decision to facilitate an adequate supply within a designated period and at an economic price. Any crop or livestock product can theoretically be contracted out using any of the models; however, certain products favour specific approaches. Broadly speaking, contract farming arrangements fall into one of five models:

1. The centralised model.
2. The nucleus estate model.
3. The multipartite model.
4. The informal model.
5. The intermediary model.

Decisions by sponsors on the type of model to follow should be made on the basis of market demand, production and processing requirements and the economic and social viability of plantation versus smallholder production. Where market requirements necessitate frequent changes to the farm technology with fairly intensive farm-level support from the sponsor, the permanent organisation and maintenance of a production chain under a centralised model is vital. Organisations that require stringent processing standards rely largely on the centralised model.

For crops such as tea, sugar and oil palm, with which farmers may have had little or no experience, sponsors are more likely to follow, where possible, the nucleus estate approach. Such crops require a significant long-term investment and, generally, immediate processing after harvest. However, the lack of adequate land or political opposition to estate development may dictate a centralised rather than nucleus estate approach. Where quality control is not the predominant concern, the informal model may suffice. In some examples, sponsors use third parties or intermediaries to subcontract production out to farmers.

If the sponsor considers that a field trial is warranted prior to the introduction of a crop to farmers or that a guaranteed minimum throughput is required for the processing facility, a nucleus estate model is often most appropriate. Where capital investment in processing facilities is considerable and the number of contract farmers is high, either the centralised or the nucleus estate structures can be used, accompanied by strong managerial inputs and backed by formal contracts. The informal model, which may become more widespread in the future, is characterised by seasonal, short-term crops with only minimal material support to farmers.

Often, the operational structure of projects changes over time. For example, the distinctions between the centralised model and the informal model are sometimes blurred. Successful individual informal developers may expand their operations into activities that eventually evolve into the centralised category. One successful small-scale developer in Indonesia started a small operation in 1970 with a few greenhouses. By 1996 the company had grown into a $US6.4 million business supplying fresh vegetables to local supermarkets and frozen vegetables for export, with the produce originating from hundreds of contracted farmers.

The Centralised Model

This is a vertically coordinated model where the sponsor purchases the crop from farmers and processes or packages and markets the product. Except in a limited number of cases, farmer quotas are normally distributed at the beginning of each growing season and quality is tightly controlled. A sponsor may purchase from tens of thousands of small-scale farmers within a single project. The centralised scheme is generally associated with tobacco, cotton, sugar cane and bananas and with tree crops such as coffee, tea, cocoa and rubber, but can also be used for poultry, pork and dairy production. Where fresh vegetables and fruits are grown under contract, the term "processing" may include grading, sorting and packaging as well as the provision of cool storage facilities.

In Africa, the contracting out of crops to farmers under centralised structures is common. These are often called "outgrower" schemes. For example, in Zambia the multinational corporation, Lonhro, considered the system preferable to growing cotton on a plantation basis. In the late 1980s it initiated a smallholder project where over 15 000 farmers grew cotton under contract for the company's ginnery.

The level of involvement of the sponsor in production can vary from a minimum where, perhaps, only the correct type of seed is provided, to the opposite extreme where the company provides land preparation, seedlings, agrochemicals and even harvesting services. The extent of the sponsor's involvement in production is rarely fixed and may depend on its requirements at a particular time or its financial circumstances. In India, a tomato processing factory in the Punjab was transferred in 1997 from one multinational company to another. The previous owners had supplied seed, supervised production and harvesting operations and provided technical advice when needed, but the new owners only provided seeds. In the Philippines, a vegetable canning company operating close to Manila decided to cease advancing fertiliser and chemicals to its contract farmers because these were being diverted to other crops and farmers were also making extra-contractual sales. The company changed to a policy of supplying only seeds unless it was convinced of the farmer's honesty.

The Nucleus Estate Model

Nucleus estates are a variation of the centralised model. In this case the sponsor of the project also owns and manages an estate plantation, which is usually close to the processing plant. The estate is often fairly large in order to provide some guarantee of throughput for the plant, but on occasion it can be relatively small, primarily serving as a trial and demonstration farm. The British-based Commonwealth Development Corporation (CDC) was a pioneer of the nucleus estate model although it no longer develops such estates. A common approach is for the sponsors to commence with a pilot estate then, after a trial period, introduce to farmers the technology and management techniques of the particular crop. Nucleus estates have often been used in connection with resettlement or transmigration schemes, such as in Indonesia and Papua New Guinea, for oil palm and other crops. While mainly used for tree crops, there are examples of the nucleus estate concept with other products. Indonesia, for example, has seen the operation of dairy nucleus estates, with the central estate being primarily used for the rearing of "parent stock".

The Multipartite Model

The multipartite model usually involves statutory bodies and private companies jointly participating with farmers. Multipartite contract farming may have separate organisations responsible for credit provision, production, management, processing and marketing. In Mexico, Kenya, and West Africa, among other countries, governments have actively invested in contract farming through joint ventures with the private sector. Multipartite structures are common in China where government departments as well as township committees and, at times, foreign companies have jointly entered into contracts with village committees and, since the early 1980s, individual farmers.

In this particular case, the county branches, through their agronomists and field technicians, were responsible for implementing and maintaining the terms and specifications of the agreement. There were formal contracts between the joint venture and the branches, and written contracts between the counties and the village committees, but only a verbal understanding between farmers and their respective committees. In theory, farmers were expected to carry out cultivation as specified by the joint venture. In practice, however, county officials only followed instructions from the joint venture if to do so was in the county branch's immediate economic interest, irrespective of quality standards and long-term production objectives. The lack of coordination between the joint venture and the county management, village cadres and farmers eventually resulted in the collapse of the venture.

In Colombia, a company started buying passion fruit in 1987, using the centralised model. The company ran into difficulties, however, because it proved impossible to control extra-contractual marketing. It therefore developed a multipartite model in which all farmers were expected to belong to associations or cooperatives, and public institutions became involved as providers of credit and extension. This arrangement significantly reduced both the risk of extra-contractual marketing and the company's costs of dealing with individual farmers, while being generally welcomed by farmers. Problems remained, however, most notably in relation to the lack of management skills on the part of the farmer associations and cooperatives.

The Informal Model

This model applies to individual entrepreneurs or small companies who normally make simple, informal production contracts with farmers on a seasonal basis, particularly for crops such as fresh vegetables, watermelons and tropical fruits. Crops usually require only a minimal amount of processing. Material inputs are often restricted to the provision of seeds and basic fertilisers, with technical advice limited to grading and quality control matters.

A common example of the informal model is where the sponsor, after purchasing the crop, simply grades and packages it for resale to the retail trade. Supermarkets frequently purchase fresh produce through individual developers and, in some cases, directly from farmers. Financial investment by such developers is usually minimal. This is the most transient and speculative of all contract farming models, with a risk of default by both the promoter and the farmer. Nevertheless, in many developing countries such developers are long established and in numerous cases they have proved an alternative to the corporate or state agency approach.

Table 2. Characteristics of contract farming structures

Structure-model	*Sponsors*	*General Characteristics*
Centralised	Private corporate sector State development agencies	Directed contract farming. Popular in many developing countries for high-value crops. Commitment to provide material and management inputs to farmers.
Nucleus estate	State development agencies Private/ public plantations Private corporate sector	Directed contract farming. Recommended for tree crops, e.g. oil palm, where technical transfer through demonstration is required. Popular for resettlement schemes. Commitment to provide material and management inputs to farmers.
Multipartite	Sponsorship by various organisations, e.g. -State development agencies -State marketing authorities -Private corporate sector - Landowners - Farmer cooperatives	Common joint-venture approach. Unless excellent coordination between sponsors, internal management difficulties likely. Usually, contract commitment to provide material and management inputs to farmers.
Informal developer	Entrepreneurs Small companies Farmer cooperatives	Not usually directed farming. Common for short-term crops; i.e. fresh vegetables to wholesalers or supermarkets. Normally minimal processing

		and few inputs to farmers. Contracts on an informal registration or verbal basis. Transitory in nature.
Intermediary (tripartite)	Private corporate sector State development agencies	Sponsors are usually from the private sector. Sponsor control of material and technical inputs varies widely. At time sponsors are unaware of the practice when illegally carried out by large-scale farmers. Can have negative consequences.

The success of informal initiatives depends on the availability of supporting services, which, in most cases, are likely to be provided by government agencies. For example, while companies following the centralised model will probably employ their own extension staff, individual developers usually have to depend on government extension services. In addition, individual developers often have limited funds to finance inputs for farmers and therefore may have to develop arrangements whereby financial institutions provide loans to farmers against the security of an agreement with the developer (an informal multipartite arrangement).

Furthermore, while nucleus estates and centralised developers frequently purchase products for which there is no other market (oil palm, tea and sugar, which depend on the availability of nearby processing facilities, or fruits and vegetables for export), individual developers often purchase crops for which there are numerous other market outlets. It is therefore important that agreements reached between the developers and farmers are backed up by law even if, in many countries, the slowness and inefficiency of the legal system make the threat of legal action over small sums a rather empty one.

In some parts of the world traders, who may not own processing or packaging facilities themselves, purchase crops for onwards sale to processors and packers. In some cases such traders provide seeds and fertiliser to the farmers with whom they deal. These are usually very informal arrangements with a high risk of default by farmers. However, in many countries, particularly in Africa, liberalisation of the export market sector has led to a breakdown of input supply arrangements in recent years and further development of such informal contractual arrangements would thus appear to merit encouragement.

The Intermediary Model

Throughout Southeast Asia the formal subcontracting of crops to intermediaries is a common practice. In Thailand, for example, large food processing companies and fresh vegetable entrepreneurs purchase crops from individual "collectors" or from farmer committees, who have their own informal arrangements with farmers. In Indonesia, this practice is widespread and is termed plasma.

The use of intermediaries must always be approached with caution because of the danger of sponsors losing control over production and over prices paid to farmers by middlemen. In addition, the technical policies and management inputs of the sponsors can become diluted and production data distorted. In short, subcontracting disconnects the direct link between the sponsor and farmer. This can result in lower income for the farmer, poorer quality standards and irregular production.

2

Specifications of Contract Farming

In most cases, agreements are made between the sponsor and the farmer, although in the case of multipartite arrangements and some others, the contracts are often between the sponsor and farmer associations or cooperatives. In the case of arrangements through intermediaries, the sponsor contracts directly with the intermediaries who make their own arrangements with farmers. Four aspects need to be considered when drafting contracts:

1. *The legal framework:* The formal law of contract in a particular country, as well as the manner in which that law is used and applied in common practice.
2. *The formula:* The clarification of the managerial responsibilities, the pricing structures and the set of technical specifications that directly regulate production.
3. *The format:* The manner in which the contract is presented.
4. *The specifications:* The details of the implementation of the contract.

The type of contract used depends on a number of factors such as the nature of the product, the primary processing required, if any, and the demands of the market in terms of supply reliability. The nature of the agreement is also influenced by quality incentives, payment arrangements, the level of control the sponsor wants to have over the production process, and the extent to which the parties have capital tied up in the contract.

A contract covering, for example, oil palm, tea or sugar, where significant long-term investment is required from all parties, will be different from a contract covering annual crops such as fruits and vegetables. A contract covering the production of fruits and vegetables for local supermarkets may not be the same as one covering such produce destined for overseas markets, which may have more rigid controls on pesticide use and product quality as well as higher presentation and packaging standards.

Although corporate bodies, government agencies and individual developers are of necessity the catalysts of the contract, farmers and their representatives must be given the opportunity to contribute to the drafting of the agreement and assist in the wording of specifications in terms farmers can understand. Any contract, however brief or informal, should represent a mutual understanding between the contracting parties. Management must ensure that agreements are fully understood by all farmers. In many countries a high proportion of farmers may be illiterate and, therefore, it may be necessary to rely on oral rather than written contracts. However, the terms and conditions entered into must be written down for independent examination and copies given to the farmers' representatives. Copies should also be available to relevant government agencies.

Legal Framework

All countries have basic laws that govern contracts. Farming contracts, whether written or oral, should comply with the minimum legal requirements that apply in a particular country. At the same time, however, it is important to take into account prevailing practices and societal attitudes towards contractual obligations, because in almost all societies these factors can produce an outcome that differs from the formal letter of the law. In some societies, for example, there may be an underlying assumption that contracts are intended to be respected only if certain factors remain constant. If climatic, political or personnel conditions should change, it may be considered socially acceptable for either party to disregard the contract, whatever the contract itself or the law may say on the subject.

Local practice may also influence the decision as to how detailed a contract should be, or whether it should be a formal contract or a more simple registration. Although there are examples of formal legal contracts that cover every eventuality, many contract farming arrangements, particularly in the developing world, are based on informal registrations. The Fiji Sugar Corporation, for example, has agreements with over 20,000 sugar-cane growers that are based on a comprehensive, legally binding document. Conversely, large tobacco schemes on the islands of Lombok and Java in Indonesia, involving tens of thousands of farmers, rely only on verbal understandings between farmers and their sponsors. In Thailand's sugar industry farmers and government personnel report never having seen a formal sugar-cane contract.

In the majority of cases, it is highly unlikely that a sponsor will take legal action against a smallholder for a breach of contract. The costs involved are inclined to be far in excess of the amount claimed, and legal action threatens the relationship between the sponsor and all farmers, not just those against whom action is being taken. Action by a farmer against a sponsor is similarly improbable. However, the improbability that a contract or agreement will be used as the basis for legal action does not mean that contracts or formal agreements should not be used. They can benefit both parties by clearly spelling out the rules of the relationship.

As neither side is likely to seek a legal remedy through the courts, it is important that ways of resolving disputes are identified in the agreement. A body representing the sponsor, farmers and other interested parties could be established in some cases, while in others a government agency might be the most appropriate forum. It is preferable that the contract farming industry regulates itself in order to offer a measure of protection for all participants. Participation of political nominees in such bodies should generally be avoided.

Agreements between sponsors and contracted farmers are essentially voluntary undertakings and, in most cases, the two parties should control their own contract formulas and specifications. In some cases, however, there are advantages in having a single body managed by the industry to regulate a number of contract farming arrangements for the same commodity. The Fijian ginger and Tongan squash industry associations were established for that very purpose.

Formula

Each contract is designed for a specific situation, the formula of which may be based on one, or a combination, of the following:

— market specifications;
— resource specifications;
— management and income specifications; or
— land ownership and land tenure specifications.

Market Specifications

Under a market specification contract only quality standards are specified. The sponsors normally provide only minimal material and technological inputs. This is the most elementary type of contract formula and is commonly used by individual developers under the informal model.

Resource Specifications

In this type of contract key components are stipulated, such as varieties and, perhaps, fertiliser rates, crop husbandry practices and the conditions under which the crop is purchased. Normally few financial or material advances are provided under resource specification formulas. Product prices tend to be based on the open market and income guarantees are minimal. Many well-established individual developers operate under the informal model and some centralised processors use this type of formula.

Management and Income Specifications

Contracts that focus on management and income specifications usually strictly regulate product standards. They are basically a combination of the market and resource specification formulas but, in addition, sponsors may establish predetermined pricing structures and make heavy commitments in the form of farm input advances, technical inputs and managerial control. This formula is the most commonly used by the multipartite, centralised and nucleus estate models.

Table 1. Characteristics of contract formulas

Type of Formula	*Sponsors*	*General Characteristics*
Market specifications	Individual developers State marketing authorities Farmer cooperatives Subcontractors	Basic controls for quality standards. Minimal inputs and conditions. Payments to farmers generally based on the open market.
Resource specifications	Individual developers Private corporate sector State agencies Farmer cooperatives	Important crop husbandry requirements in contract. Payments to farmers sometimes based on open market, sometimes fixed. Limited material inputs to farmers.
Management and income specifications	Private corporate sector National development agencies	Directed contract farming. Intense contract formulas; high degree of material and management inputs. Prices to farmers are fixed and adjusted on a seasonal basis. Common under centralised and nucleus estate models.
Land tenure specifications	Private corporate sector National development agencies	Clauses relating to land tenure and land use conditions part of the formula. Normally directed contract farming. Intense contract formulas, high degree of material and management inputs. Common under centralised and nucleus estate models.

Land Ownership and Land Tenure Specifications

This type of contract is an extension of the management and income model, with additional clauses relating to land tenure. Wherever private companies or government agencies lease land to farmers for contract farming, formal crop-land tenancy contracts on a long-term basis are necessary. These contracts should be legally binding and can contain clauses relating to both crop and land husbandry. Land tenure specifications may stipulate the other crops that farmers are allowed to cultivate in proximity to the contracted crop. This is in order to avoid the risk of disease.

One early example of a land tenure contract was the Gezira cotton venture in the Sudan referred to in the Introduction. In spite of its "colonial" nature the Gezira project helped introduce an innovative concept of agricultural development. The venture established control over the use of the land, insisted on terms of tenancy related to an economically viable unit and developed sound agricultural rotation. Such measures led to production efficiencies previously only attainable by large-scale estate management.

One land tenancy agreement for a cotton contract scheme in Zimbabwe stipulated that there was an "explicit restriction on non-farm activities." The restriction centred on the objective of the sponsors to confine farmers to growing only cotton. Formal tenancies of this nature should at least allow for subsistence production for farmers and their families.

Many contract farming ventures are in areas where customary land usage arrangements are negotiated between landless farmers and traditional landowners. While this allows the poorest cultivator to be included, sensitive measures must be applied to ensure that contracted farmers are not exploited as a result of landowners charging excessive rents.

Formats

The various formats that a contract may take are:

- — formal agreements;
- — simple registrations; and
- — verbal agreements.

Formal Agreements

Explicit, legally endorsed contract formats, which closely detail the conditions and obligations of each party, are particularly common in projects that involve heavy investment in capital infrastructure or where sponsors lease land to farmers specifically to grow crops under contract. However, such contracts can also be used when land tenure is not a factor.

Simple Registrations

These are a common contract format used by most centralised operations and, to a lesser degree, under the informal model. The term "registration" usually refers to a signed confirmation from the farmer that he/she wishes the sponsor to reserve a contract for him/her. Simple registrations are based on so-called "informal associations of trust and patronage that bypass formal legalities." With a flexible and sensitive managerial approach, a simple registration is a proven and practical way to sustain contractual arrangements. The technical aspects of the agreement are drafted in short, simple terms, clarifying the responsibilities of both sponsor and farmer.

Normally, the registration of farmers for the following season commences immediately after the last harvest. In well-established projects, registration for many farmers is only a formality or, perhaps, involves just a change of name of a family member. Following registration, field staff approve the land on which the crop is to be cultivated and decide on production quotas based on potential performance.

Verbal Agreements

Unwritten or verbal agreements are commonly used by informal individual developers and sometimes by corporate sponsors. A major problem of verbal agreements is the interpretation of responsibilities and specifications. Confusion and misunderstanding can easily occur if the agreements are not clearly explained by management to the farmers and their representatives. In turn, the managers' field extension staff must also have a clear understanding of the terms of the agreement.

Specifications

Contracts will need to specify some or all of the following aspects of the sponsor-farmer agreement:

- — contract duration;
- — quality standards;
- — production quotas;
- — cultivation practices;
- — crop delivery arrangements;
- — pricing arrangements;
- — payment procedures; and
- — insurance arrangements.

Contract Duration

The duration of agreements depends on the nature of the crop. Contracts for short-term crops such as table vegetables are normally issued and renegotiated on a seasonal basis, whereas crops such as tea, coffee, sugar cane, and cocoa require long-term contracts that can be amended periodically.

Quality Standards

Product quality or, more precisely, the absence of quality, can have far-reaching consequences in terms of market acceptance and future expansion. Most contracts contain detailed quality specifications so that produce that does not conform to the agreed criteria can be rejected. It is important that farmers fully understand the reasons for standards and also understand that the acceptance of poor quality produce from some farmers will ultimately affect an entire project and thus there is no long-term advantage to individual farmers to try to cheat. In the case of most smallholder tea schemes, for example, all leaf purchased on one day is processed at the same time. Poor quality green tea delivered by a few farmers will reduce the overall quality of the processed tea, thereby reducing returns for all.

Quality specifications may specify the size and weight of the product, the degree of maturity and the manner in which it is packaged and presented. A major problem with quality standards is that they are frequently vague and not clearly understood. This uncertainty could cause corruption problems, for instance the sponsors' employees seeking bribes to upgrade produce, or unfair practices by management such as trying to downgrade produce in order to reduce purchases when market conditions are poor.

The use of terms such as "grade one, grade two," or "first quality, second quality" without clear specifications as to what these mean is unacceptable. The description of each grade is kept as simple as possible, yet distinctly highlights the grade parameters. Extension staff should demonstrate the grades to farmers at the beginning of each season and explain the rationale for the specifications. It may also be necessary to specify the maximum content of each delivery container as quality can suffer if containers are overpacked.

Wherever possible, the number of grades should be kept to a minimum and each grade's specifications should be presented in clear terms. Unfortunately some contracts have demanded a complex multiple-grade system. In one instance, for example, there were 41 different grading specifications in a single contract. This resulted in widespread confusion that led to misunderstandings and, ultimately, confrontation. In another case strike action by farmers on a project in the South Pacific continued for two seasons and ended only when management introduced a simplified grading system.

The grades were reduced from twenty-nine to three, thus making things more practical for the farmers. There was no subsequent deterioration in crop quality because of the introduction of a new processing technique designed to accommodate the simplification of the grading system. Often, however, there may be a need for only one standard, with all produce delivered being required to fall within a particular specification range. For example, in the case of papaya for export from Fiji only a single price was offered. The grade specifications were based solely on the size and maturity of a single variety, these two components being critical for market acceptance.

Raw materials for processing are often purchased on the basis of the likely extraction rate. For example, sugar-cane deliveries are often sampled to ensure that they meet the minimum juice purity specified in the contract. In France and Italy, among other countries, members of wine cooperatives are paid according to the sugar content of their grapes. Such individual calculations, however, are not usually possible with relatively small-scale farmers as their product is bulked up with that of several others for transport to the processing facility. Oil palm farmers, for example, are usually paid on the basis of the average "fresh fruit bunch" conversion rate achieved by the factory over a specified period.

Production Quotas

Both insufficient and excessive production can have serious ramifications. Overproduction can mean unpopular quota reductions and costly stockpiles. Conversely, underproduction caused by poor farmer selection, disease or climatic factors could eventually result in a project becoming insolvent, as processing costs per tonne could rise to unacceptable levels. Moreover, if a processing plant is unable to meet pre-arranged marketing contracts, future orders could be decreased or cancelled. Quotas are employed in the majority of contracts in order to:

- — utilise processing, storage and marketing capacities efficiently;
- — guarantee markets for all farmers;
- — ensure quality control; and
- — monitor farmers' performance.

Utilise processing, storage and marketing capacities efficiently: Failure to purchase any part of a farmer's production that meets the specifications of the contract will cause serious discontent. This could occur through poor calculation of the sponsor's capacity to handle, process, store and market the production. Sponsors need to limit their contractual commitment to buy from farmers to only the quantities they can process (in total and, depending on the product, on any one working day) and market. Processing is also often restricted by the capacities of the sponsors' and farmers' warehouses.

Quotas allocated on the basis of actual volume to be produced by each farmer or, alternatively, on a defined area to be planted should overcome some of these concerns. In Thailand, for example, cassava farmers are issued rootstock of a specific variety by their sponsor. The quantity of the stock issued is proportional to the area that the extension staff considers sufficient for the farmer to transplant, plant and cultivate. The advance of the rootstock material becomes, therefore, the production quota. All production from the allocated rootstock is purchased in accordance with the grading specifications that indicate the specific gravity and the degree of pest damage of the cassava.

Guarantee markets for all farmers: Specific quotas allocated by managers in each individual contract should guarantee that all farmers would be able to sell all of their production that meets the agreement's conditions. Without quotas, sales by farmers would be on a "first-come-first-served" basis. Theoretically, deliveries by some farmers could exceed the requirements of the sponsors, resulting in some other farmers being unable to sell any or only part of their crop.

Ensure quality control: Quotas can sometimes be used to control the quality of the raw commodity. For some crops, increases in quantity may only be achieved as a result of reduced quality, for example, as measured by the extraction rate. Farmers may be tempted to manipulate weights through fraudulent methods such as wetting the crop before sale or by adding foreign matter. Sponsors can reduce the likelihood of such practices by setting quotas based on expected output for a given area.

Monitor farmers' performance: As a means of monitoring farmers' production, quotas are used as a benchmark by which to analyse crop yields. Through the efficient monitoring of crop development, extension officers can usually make realistic yield forecasts. If yields fluctuate widely then abnormalities can be investigated and remedial measures taken. The use of quotas also permits the sponsor to identify whether farmers are selling crops outside of the contract (i.e. extra-contractual marketing) or whether they are supplementing their sales to the sponsor with non-contracted production from other farmers.

The allocation and distribution of production quotas will vary according to crop and circumstances. Where there is no alternative market for the crop and farmers have made significant long-term investments in production (tree crops) or processing facilities (e.g. tobacco curing barns), the sponsor must be committed to purchase the entire crop covered by the quota. This obligation, of course, is subject to the crop meeting the agreed quality specifications. The most common and practical method is to allocate quotas on an area basis, with managers calculating the total area to be cultivated in relation to the project's processing capacity and their knowledge of each farmer's expected yield.

Managers must also address the issue of how the quotas will be allocated between locations and between the farmers they select. In Mozambique's Nampula province, for example, sponsors assign a similar land area to all farmers included in the schemes. While farmers can be rejected for failing to fulfil production targets or repay credit, land size allocations do not vary. Such an approach is seen as avoiding conflict within the community and avoiding corruption in quota allocations.

On the other hand, allocating a quota that the farmer cannot cultivate, either by area or quantity, will cause serious problems. Reductions of quotas in subsequent years or cancellation of contracts on the grounds of failure to supply agreed quantities could cause demoralisation and a loss of prestige for the farmer. The allocation of appropriate quotas that reflect the different levels of resources and skills of the farmers and, at the same time, permit a wide range of farmers to have contracts will add to the stability of contract farming ventures.

Where there are alternative markets for crops under contract, quite often farmers are tempted to sell outside the contract. Quotas deliberately set at levels lower than the farmers' actual production capacity may enable them to take advantage of high open market prices when they occur. Such an arrangement is likely to apply particularly when the pricing arrangement is for a fixed price rather than a market-based price. An example of this is found in passion fruit production in Colombia. Such an arrangement cannot be acceptable in all cases; for example, a company that supplies farmers with day-old chicks and their feed will obviously expect the farmers to deliver the same number of chickens minus, of course, deaths.

There are examples of contracts where quotas are superfluous. This particularly applies when the crop has several buyers in active competition, and when it is unlikely that the company will be able to purchase more than the throughput of its processing facilities and thus wants to maximise its purchases. It is most common under the informal model, an example being the cotton industries of Zambia and Zimbabwe where several gins are in active competition for the available crop. In other circumstances, however, the absence of quotas can work to the disadvantage of the farmer. For example, although farmers in the Punjab grow tomatoes under formal contracts, the sponsors do not issue fixed quotas. The sponsors purchase only the amount they require leaving farmers with no option but to sell surpluses on the open market at a reduced price. The buyer therefore has much greater bargaining power than the farmers.

The sale of quotas between farmers should be discouraged in that the new farmer may not meet the farmer-selection criteria and the practice could lead to corruption. In all ventures there is always a degree of attrition; some farmers die, others retire or sell their land and move to other districts. In such situations, management usually transfers the contract to family members or nominees of the previous contractor on the understanding that the newcomer meets the selection conditions.

Cultivation Practices

When sponsors provide seeds, fertilisers and agrochemicals, they have the right to expect that those inputs will be used in the correct quantities. They also have the right to expect that farmers follow the recommended cultivation practices. Of particular concern is the possibility that farmers may apply unauthorised or illegal agrochemicals which can result in toxic residues, with dramatic repercussions for market sales. It is therefore essential that all contracted farmers adhere strictly to the project's input policies. Managers and their extension staff must make every effort to explain to farmers why the specifications and input recommendations must be followed.

Crop Delivery Arrangements

Arrangements for collection of products or delivery by the farmers vary widely. Some ventures stipulate that farmers should deliver their harvest to processing plants at given dates; others may include the use of the sponsor's transport to collect harvested crops at centrally located buying points. For contracted fresh vegetables a normal practice is farmgate collection. When the sponsor's transport is used there is normally no cost to the farmer.

In the sugar industry, small railways are used extensively; farmers deliver their harvested cane to a central loading point from which it is then transported to the crushing mill, weighed and purchased. Many formal contracts have clauses that outline the obligations of both the farmer and the sponsor regarding delivery and collection respectively. As a routine practice, managers and their extension staff should confirm delivery or collection arrangements at the beginning of each season and reconfirm these prior to harvest.

Pricing and Payment Arrangements

Pricing and payment arrangements are the most challenging components of all farming contracts. The choice of which crop pricing structure to use is influenced by whether the crop is for the local or export market, the seasonal nature of production and the degree of competition in the marketing system. The application of transparent pricing formulas is crucial and the drafting of a clear pricing structure and the organisation of a practical method of payment encourage confidence and goodwill. There are several ways prices offered to farmers can be calculated, including:

— fixed prices;
— flexible prices;
— prices calculated on spot-market values;
— prices on a consignment basis; and
— split pricing.

Fixed prices: Fixed prices are the most common method. The practice is usually to offer farmers set prices at the beginning of each season. In almost all cases, fixed prices are related to grade specifications. In calculating prices there may be a tendency for sponsors to adopt a cautious approach because of the danger of market price fluctuations. Fixed price formulas are usually ideal for the sponsor; however, where alternative outlets exist, farmers may consider such arrangements to be disadvantageous if prices increase on the open market.

For managers, the set price formulas are preferable for both budgeting and marketing purposes, although they are still obliged to purchase the crop at the prices stipulated in the contract if the open market prices decrease below the set prices. The fixed price structure is widely used by tobacco corporations and companies processing crops for canning.

Flexible prices: This structure applies to prices calculated on a formula related to changing global and local markets. This form of pricing is common in, for example, the sugar industry where the final price to the farmer is known only after the processed sugar has been sold. Farmers are paid on the basis of a formula which takes into account agreed processing and other costs of the sponsor as well as world market prices over a particular period. The prices of internationally traded commodities for which there are few, if any, grades are readily accessible and should also be made available to farmers. In Papua New Guinea, smallholder oil-palm producers on nucleus estates are paid on the basis of such a formula, which is monitored and approved by the Government. In Guyana, sugar-cane producers receive two-thirds of returns from sugar sales and the factory one-third, to cover costs and profit.

In some cases farmers and sponsors may share price increases and costs proportionately. In the Philippines, for example, a pig-rearing contract specifies that the farmers and the company shall divide proceeds equally, after deduction of the agreed expenses of the company. The expenses include stock feed, medication, a marketing fee and an allowance for shrinkage between delivery of the animal and eventual sale. Such a profit-sharing arrangement can be successful if the contractor is efficient and honest. However, in other circumstances this system can seriously prejudice farmers by putting them at the mercy of inefficient processing and marketing. Wherever payments are dependent on fluctuating markets an independent arbitration mechanism should be developed by the industry to safeguard the interests of both the farmers and the sponsors.

Prices calculated on spot-market values. Payments based on spot-market prices can be very complex and often lead to misunderstandings and disputes. Such an arrangement removes income guarantees for farmers but does enable them to take full advantage of high market prices. The main problem with this approach is that sponsors and farmers must arrive at a common understanding of what constitutes a market price that is relevant to the higher quality that contracted farmers could be expected to produce. This form of pricing is common in Thailand where individual small-scale developers act as brokers under informal contracts. The brokers make arrangements with farmer groups to sell fresh vegetables to wholesalers. They collect the crops at the farmgate, arrange transport to Bangkok and, after the produce has been sold, pay the farmers a percentage of the final sale price. In most cases the open market pricing system is unsatisfactory, as the farmers do not have control over the price they receive or knowledge of how it is calculated.

Prices on a consignment basis: Prices calculated after the produce has been marketed and sold may be considered another form of spot-market pricing. This form of payment is normally termed "on consignment" and is mainly used by informal small-scale developers. In another example from Thailand, individual developers arrange to supply crops to markets on consignment. They take a commission out of the farmers' revenue and, at the same time, deduct the costs of seed and fertiliser advanced to the farmers. Consignment pricing arrangements are rarely found in well-structured contract farming projects and are best avoided. The growing importance of supermarkets suggests that more and more fresh produce will be delivered at predetermined prices rather than on a consignment basis.

Split pricing: Under this system an agreed base price is paid out at the time of purchase or at the end of the harvesting season. The final price is calculated once the sponsor has on-sold the commodity, and depends on the prevailing market price. If the crop is sold in the fresh form the second price can usually be calculated within a month. When the product is processed it may take much longer.

Payment Procedures

For all farmers the most convenient method of payment is usually cash-in-hand immediately following delivery of any part of their crops. However, this is not always possible, particularly if the sponsor has limited resources, where payment depends on the total production after processing, or where the payment is based on the price the sponsor obtains. Also, the company may have the obligation to repay loans advanced by banks to farmers using the contract as collateral. In the majority of cases payments are made periodically throughout a season, perhaps two to four times, with the final payment after the last harvest. Any material and cash advances given to farmers during the season are normally deducted from the final payment.

Insurance Arrangements

Agricultural investments always involve risk. The five most likely reasons for investment failure are poor crop management, climatic calamities, pest epidemics, market collapse and price fluctuations. The standard agribusiness approach to indemnify against quantity shortfalls is crop insurance. Although government-run crop insurance schemes often prove to be unsatisfactory, success with insurance programmes offering named (i.e. limited) peril cover for certain crops has been reported in Mauritius, the Philippines and Cyprus. There is also a growing awareness by the private sector that crop insurance should be encouraged and promoted. As the farming involved in a contract arrangement becomes technologically more advanced, the range of risks to which it is subject generally becomes more limited. In many cases some of the remaining risks can be managed with the assistance of insurance.

For seasonal contracts that are based on a fixed price payable at harvest, it is unlikely that farmers will be affected by market collapse or price variations; however, they are vulnerable to production losses caused by climatic or human factors. Some authorities classify the four main categories of crop insurance in order of "their comprehensiveness in terms of coverage of risks" as follows :

— "Acts of God";

— destruction of specified assets;

— loan default; and

— production and income loss.

"Acts of God": This category encompasses natural disasters such as drought, floods, hail, storms, cyclones, lightning, insect plagues and disease epidemics that are beyond management control. In India, for example, efforts have been made to introduce rain-insurance contracts based on rainfall duration, the precise rainfall shortfall and a prearranged schedule of indemnities. The degree of insurance compensation for damage by flood, storms or hail is difficult to assess when there is only partial loss. Field crops that have been damaged by a fractional flood or hail require an experienced and independent assessor who would not only have to evaluate quantity losses but also the quality of the crop at the time of the damage. Some contracts may have "Acts of God" clauses inserted, although such clauses are uncommon.

Destruction of specified assets: Many farmers insure their houses, garages, grading and storage sheds as a matter of routine. Tractors and farm implements can be insured against damage and theft. Insuring of curing kilns by farmers when growing tobacco under contract is essential. If a contract farming venture is well established, management can sometimes organise the insurance of non-contractual farm buildings and housing as part of the sponsor's total indemnity policies, reducing the cost of premiums to the grower.

Loan default: In almost all ventures, sponsors assume the liability of credits advanced by management to the farmer for the contracted crop. It is therefore important that advances do not accumulate into debts that the farmer cannot repay. Managers normally allow farmers who cannot repay advances because of climatic or other mitigating factors to extend their loans to the following seasons. Sponsors, of course, do have the option to indemnify their farmer loans against default through their own insurance brokers.

Production and income loss: Insurance against both production and income loss is expensive and complex. Production loss may be caused by a combination of factors that are difficult to insure against. To determine who is culpable when a crop is destroyed by insects is one example. Was it an "Act of God" or the failure of the farmer to take measures for pest control at the appropriate time, or was it the fault of management for not training and instructing the farmers in pesticide techniques? There are also social risks that could cause crop loss such as theft and animal damage. In the case of injury to crops by horses, goats or cattle, the project's management staff assess the value of the destroyed crop. If the owners of the animals are not contracted farmers, local community leaders can sometimes negotiate compensation. If the owners are also contracted farmers, compensation disputes can often be negotiated through farmer forums.

Where there are fixed price contracts there is no apparent risk to farmers with regard to payment for their crops. If a market collapses, the sponsor should automatically shoulder the loss. However, if the sponsor becomes bankrupt, farmers could be permanently affected. Where contracts are on a flexible or spot-price basis the stability of farmers' incomes is always at risk. In theory, the proposal of crop and property insurance for farmers in contract farming ventures is appealing. However, before advising farmers to consider insurance, a qualified risk analysis should be made to determinate the economic advantages of insurance against the specific risks applicable to the particular crop.

3

Managing the Contract Farming

Poor management can cause potentially promising contract farming ventures to fail. This chapter describes the steps that managers must take in order to coordinate production activities and the delivery of products by farmers to the processing and/or marketing facilities.

Coordination of Production

A number of specific organisational and administrative activities have to be carried out before production commences. The key issues that managers must address in advance are:

- — identifying suitable production areas;
- — selecting farmers;
- — forming working groups;
- — providing material inputs;
- — providing logistical support; and
- — purchasing the product.

Identifying Suitable Production Areas

Following the choice of a suitable physical and social environment, management must then select specific areas that can provide easy access for extension and logistical services. It is usually important for the contracted farmers to be situated relatively close to the company's processing or packing facility. Not only does this allow perishable produce to be processed without delay but it also permits management and technical and extension staff to be based at one centralised location. The raw material for many crops is far more bulky than the finished product, and by locating the processing facilities closer to the primary producers transport costs can be minimised.

There are, however, situations where it is more practical to spread production over a larger area. This may be when farmers are situated in many districts and channel their production to a number of processing units controlled by a single sponsor. The cultivation of crops in different areas can also reduce the risk of total crop failure due to irregular weather patterns or disease. Such a strategy helps to guarantee a regular supply for processing and marketing. In such cases it is more convenient for extension officers to reside in or near their areas of responsibility. At times the most advantageous areas for growing the crops may not be the most logical locations for a factory.

Vegetables for canning, for example, can sometimes be grown long distances from the factory. A canner in the Philippines organised farmers situated over a wide area to grow both sweet and baby corn to take advantage of climatic variations in order to maintain regular supplies throughout the year. In Kenya, French beans for canning are grown between 150 and 230 kilometres from the factory because the climate in the production area allows for the beans to be grown under rain-fed conditions. Where several areas are chosen, they must all be able to provide a sufficient quantity of production to ensure that services provided to the farmers can be carried out in a cost-effective manner.

Selecting Farmers

Following the choice of production areas the next requirement is to select farmers. Management must decide how many farmers should be offered contracts and the criteria for their selection. Farmers can be approached individually, through the offices of agriculture departments, through community leaders and farming cooperatives, or by an open invitation to apply. Selection criteria should be based initially on an assessment of the suitability of the farmers' land and confirmation of their tenure security. If those two conditions are met, an evaluation needs to be made of the prospective contractor's farming experience, past production record, desire to cooperate and the extent of his/her family labour inputs. All selection appraisals must take into account the complexity of the household economy and examine how the contracted crop can be incorporated within the farmer's total farming mix.

Community leaders and local government officials are, in general, dependable sources of knowledge on the capabilities and attitudes of farmers in their villages and districts. Managers should be aware, however, that petty rivalries and extended family obligations are characteristics of some rural societies. Farmer selection therefore should also be judged on the manager's own intuition and available independent assessments. In the case of the production of French beans in Kenya, mentioned above, the factory uses local government administration, government agricultural extension offices and its own field staff to select farmers. Criteria used for selection are soil type, the agricultural experience, competence and reliability of the farmers, combined with their ability to cooperate with others.

While failure to select some farmers may cause resentment, the arbitrary selection of farmers who fail to produce the desired quality and quantities can be commercially disastrous. In one venture in Thailand, for example, farmer selection for the cultivation of vegetables for canning was deemed to be very lax. Because of high product demand and land shortages, the company accepted virtually all farmers. Furthermore, application forms were circulated after most farmers had signed their contracts, a practice that caused an atmosphere of confusion and uncertainty. Managers should, wherever possible, verify that the production potential of any district is in excess of their requirements in order to provide them flexibility to choose the most qualified farmers.

The criteria for farmer selection are likely to vary according to the type of crop. Less rigorous standards can be adopted for short-term seasonal crops in that farmers who fail to perform can be excluded from subsequent contracts. For tree crops, however, a long-term commitment is required and thus sponsors need to be assured of the reliability of the farmers and of their ability to continue to farm for many years. In an oil palm venture in Ghana, for example, the majority of the selected farmers were "veterans" with at least twenty-five years experience. This resulted in an age and marital composition that could be expected to raise constraints for future production because there were few younger farmers and the farmers were limited to immediate family labour only. Transmigration or settlement schemes in a nucleus estate context involve the risk that the farmers will be unhappy in their new environment and wish to return to their original homes. Rigorous selection procedures can minimise, but not altogether avoid, such risk.

Crops such as cotton, maize, tobacco and vegetables are grown under contracts that are normally reviewed and renegotiated on a seasonal basis. Periodic reviews allow for pricing and technical adjustments at the beginning of each season, for new farmers to be registered and, where appropriate, for the quotas of farmers who are less productive to be reduced to levels they can reasonably manage. When a farmer requires the use of outside labour, an assessment of the availability of such labour and the farmer's ability to manage it will be necessary.

Forming Working Groups

Although a company usually deals with farmers directly through its extension service, it can be advantageous to organise small-scale farmers into groups. Such groupings can serve a number of functions such as taking joint responsibility for credit or cash advances, monitoring applications for inputs and arranging for the delivery of inputs and collection of the crop after harvest. Formally organised groups can provide suitable units for delivery of extension advice. In Malawi, for example, tea growers under one contract are divided into twenty-one blocks. Each has a committee responsible for matters such as settling disputes and ensuring that members follow the recommended crop husbandry practices and abide by the conditions of the contract. Some tenant tobacco growers in Malawi are also divided into blocks, with each block having a leader who organises group activities, including a seedling nursery and soil-conservation activities. The farmers apply fertiliser to their fields as a group in order to avoid diversion of the fertiliser to food gardens or to the open market.

The largest cotton company in Zimbabwe supplies inputs to self-selecting farmer groups. The whole group is penalised if one member defaults, so there is an incentive for peer policing to ensure repayment. Monetary rewards are given to groups with high repayment rates. On the other hand, defaulters are followed up quickly and assets can be seized. The company originally established groups with a minimum of fifty farmers, but experience has shown that to be too many and it now works with a maximum of twenty-five farmers per group. A company that exports vanilla from Uganda works through groups of farmers organised into local associations. The associations play a leading role in selecting suitable farmers, recovering loans and bulking up the vanilla for purchase.

Providing Material Inputs

The provision of material inputs to farmers is an important feature of contract farming. Before the start of each season, managers should calculate the pesticide and fertiliser requirements for each farmer based on his or her production quotas. Material advances can, on occasions, extend to the financing of draught oxen and horses, ploughs, spraying equipment, small irrigation pumps and other ancillary farm tools. Tractors and heavy machinery can also be advanced to farmers who have proven performance records. Sponsors sometimes guarantee instalment payments to banks and credit agencies for heavy equipment. Such payments are deducted from the farmer's crop proceeds and, naturally, can be extended over a number of years.

All inputs should be ordered and supplied to the farmer well in advance of sowing or transplanting. The normal administrative procedure is for the administration staff to maintain a statement of account for each farmer, detailing credits of crop purchases and debits for material advances. The normal method of repayment for inputs and other assistance supplied is for the advances to be deducted prior to the final payment. Pricing of inputs and production services can be a sensitive issue and sponsors should charge no more than prevailing commercial prices. Where the sponsor buys inputs in bulk for delivery to farmers, charges made for handling and transport should be clearly explained to the farmers.

Charges to the farmers for services such as ploughing and harvesting should not be augmented to cover the inefficiency of management. When available, farmers should have the option to use commercial ploughing services if they are less expensive than those offered by the company. Where outside contractors are used, it is the responsibility of the extension staff to ensure that the work meets the required standards. The normal practice is that management pays the cost of the service to the tractor contractors after the work has been approved and then charges that cost to the farmers' accounts.

In Ghana, cotton companies used to provide "free" inputs to farmers and then offer a relatively low price for seed cotton in order to recover the cost. However, this was found unsatisfactory as, in effect, the more farmers produced the more they paid back, and the companies soon changed to the system of deducting individual farmer advances. The idea of paying a low product price to recoup the cost of inputs supplied on credit is not new. It was, for example, widely practised by export crop boards in much of Africa.

Providing Logistical Support

Another key management function is to organise input distribution, container deliveries (bags, boxes, slings, etc.) and the strict timetabling of transport, especially at harvesting time. This is a vital area for management because logistical problems jeopardise both the sponsor's profitability and the relationship between sponsor and farmer. In the Philippines, for example, the sponsor of a broiler contract experienced problems because the suppliers had, at times, difficulties in ensuring that poultry feed reached the farms on scheduled dates.

Farm managers responded by either reducing the quantity of feed, which caused cannibalism among the birds, or buying substitute feeds, which may not have contained the same formulation of ingredients and supplements. There were also problems resulting from the late collection of marketable birds due to poor transport arrangements. In addition, there were allegations of pilferage of live birds and feed by the company's agents.

There can be particular problems where a crop requires processing immediately following harvest. Such crops include tea, sugar cane, tobacco and some vegetables for canning. In Malawi, tea grown by smallholders was rejected because it became over-wilted owing to the failure of the company to provide transport immediately after harvest. The company was obliged to advise farmers of its inability to provide transport at scheduled times, but often this was not done. Although the company was negligent, no compensation was paid. A similar situation has arisen in the sugar industry in Kenya. The sugar company provides transport but it is the responsibility of the farmers to ensure that their sugar cane is delivered within seventy-two hours. Poor logistical support will inevitably sour relations with farmers and reduce the viability of projects.

Purchasing the Product

Management must ensure that staff are available to purchase the product from farmers as scheduled. Significant efforts must be made to avoid corruption in the buying process. Farmers must be able to verify the weights of the products they sell to the sponsor. Also, where produce is rejected without the farmer being present there is inevitable suspicion. Under no circumstances should the sponsor dispose of rejected produce without first giving the farmer the opportunity to inspect it. Depending on the circumstances, buying may be carried out at the farmgate, at centralised buying locations or sometimes at the processing plant.

Most produce must be purchased shortly after harvest or farm processing so that it is presented in its most favourable condition. When buying is carried out at the processing plant it is often impractical for rejected produce to be sent back to the farmer. However, the extension staff should notify farmers of the reasons why their produce was rejected and offer them the chance to visit the plant to inspect the rejected consignment. Good managers ensure that farmers or their representatives are present when produce is purchased.

Managing the Agronomy

In China a large joint venture involving over 23,000 farmers ceased to operate after seven years because of management's failure to organise and direct harvesting and grading practices. It is important, therefore, that managers not only form competent field extension teams, but also plan effective production schedules. During the production season, supervision by extension services of all cultivation activities is essential, particularly to be sure that certain recommended practices are compatible with the farmers' ability to implement them. Significant factors in any venture's performance include:

— field extension services;

— transfer of technology;

— cropping schedules; and

— training and research.

Field Extension Services

Extension staff employed in contract farming ventures are usually the key link and the direct interface between the sponsor's management and farmers. They require a number of key skills that include:

— A good comprehension of the crop(s) or animals under contract.

— A sensitive and sympathetic understanding of the social customs, language and farming practices of the communities they work with.

— An ability to communicate effectively with farmers and to organise and administer cropping schedules and buying procedures honestly and impartially.

They must also possess an understanding of agronomy, farm management techniques and the potential capabilities of the farmers with whom they work. When selecting extension personnel, consideration should be given to both the personal aptitudes and the formal qualifications of each applicant. In larger projects, senior extension staff and field agronomists should normally be educated to tertiary level. Junior personnel can be recruited from within local farming communities. Although they may lack formal education, they will have the advantage of local knowledge and field staff enlisted from local families can have a beneficial effect on sponsor-farmer relationships. Conversely, as a result of close relationships with immediate and extended families, there is a chance that quota distribution and buying practices can become corrupted.

Extension staff must first obtain the credibility and trust of the farmers they advise in order to successfully implement the policies of the sponsor. It will then be much easier for them to administer the strict regulatory procedures that are frequently necessary for the maintenance of quality standards and crop uniformity. The deployment of extension staff may vary considerably. In the Pacific, a company buying maize, tobacco, vegetable crops and export papaya allocated one field officer to every 55 farmers.

In a rice project in West Africa, each extension officer supervised 300 farmers, while a joint venture in China provided one technician for every 500 farmers. The degree of responsibility of field extension staff depends on the project's structure whereas the extension worker to farmer ratio mainly depends on the type of venture. Sugar cane and cereal crops require a lower ratio. Intensive crops that need constant supervision are cut flowers, vegetables and tobacco. In addition to their job of coordinating farmers, the biggest challenge for extension staff is probably to encourage farmer participation in decision making while maintaining positive sponsor-farmer relationships.

Transfer of Technology

Farmers will only accept new techniques if the

adaptations result in higher yields and/or improved quality and if the cost of such techniques is more than offset by higher returns. The introduction of technologies can cause cultural adaptation problems for smallholder farmers, even though these technologies are often the most important benefit of the contract. A reported example of this comes from the Central American banana industry. It was found that the potential for gaining acceptance of new technology was greatest with farmers who had little or no prior knowledge of the crop.

The new growers had no preconceived ideas on how to grow bananas so they readily accepted the new technology. Managers and agribusiness agronomists should consider the following four basic questions before transferring technology through their extension staff:

1. *Intellectual capacity:* Do their employees have the ability to transfer appropriate changes?
2. *Technical feasibility:* Do the farmers have the inclination to accept new techniques?
3. *Economic viability:* Is the adaptation too costly, complex and risky for the farmer?
4. *Cultural acceptability:* Does change affect the general farming practices of the farmer?

The process of introducing new technology normally progresses through a number of stages as follows:

1. *Awareness:* This is the stage when the concepts of adaptations are introduced by the sponsors' agronomists to their extension staff. The implications of these adaptations are, in turn, explained to farmers and their representatives.

2. *Benefit:* It is often necessary to carry out field trials in order to assess the benefits of new technology before it is introduced to farmers. Such trials should either realistically simulate conditions faced by farmers in contracted areas or be carried out by a sample of contracted farmers. Sponsors should not make sizeable investments, nor should they encourage farmers to do so unless there are strong indications that the adaptations will be profitable.
3. *Implementation:* This involves the extension of the adaptations to farmers through field demonstrations, trial plots and lectures by the technical and extension staff.
4. *Appraisal:* Appraisals of yields and quality will demonstrate to both farmers and extension officers the value of the adaptation. Such appraisals are particularly important when new varieties are introduced. Farmers who have obtained good results from a particular variety may be reluctant to plant a different one that may yield less but has other important characteristics, such as a high degree of tolerance to disease.

Cropping Schedules

Scheduling refers to the timing of specific activities and the sequence in which those activities are carried out. This is important in order to accommodate climatic patterns and to ensure regular and consistent product supply to the sponsor's processing and/or packaging facilities. Efficient scheduling is vital where production by thousands of farmers has to be synchronised for the purposes of irrigation, the achievement of uniform crop quality and the organisation of transportation at harvest. Processing capacities can be increased appreciably, for example, by implementing staggered transplanting. This may allow farmers to plant two fields in the same season, perhaps only four weeks apart.

Extension staff have the responsibility to schedule the sowing of seed beds, the transplanting of seedlings, and the cultivation and harvesting of the contracted crop within a defined climatic season and in harmony with the farmers' own cropping regimes. At the beginning of each season, management, extension staff and farmers should discuss and confirm all planned activity schedules. Managers should present the sequence and timing of each crop activity to farmers before the first sowings.

Strict regimes imposed by crop schedules can change the work routines of farmers. This can alter social relations, the gender division of labour and the control and utilisation of land and farm resources. Farmers necessarily forfeit some degree of autonomy when they accept exacting timetables and contract specifications, but they do so in expectation of greater economic rewards.

Training and Research

Management may consider organising training programmes for extension staff and farmers in the form of regular lectures and field days as well as through the use of demonstration plots. Staff training can be provided by in-house lecturers and visiting teachers from scientific institutions or through external training courses. Only when the extension officers have gained a comprehensive knowledge of the relevant product(s) and an understanding of their administrative responsibilities are they really in a position to transfer technology to farmers.

There are a number of ways in which the training of farmers can take place. Routine farm inspection visits by extension staff can usually include an element of technology transfer. More formal and regular meetings with farmer groups, conducted by senior extension and research personnel, can be held to concentrate on the relevant activity at the time, e.g. seed sowing, transplanting, fertilising, pest and disease control or harvesting. There may be a need for up to six such meetings within a single season.

When potentially controversial subjects such as grading standards and buying procedures are explained, it is preferable that project managers also attend. Another way to demonstrate innovative farming methods and improved varieties is by having field days in the sponsor's trial plots or in the fields of leading farmers. Management and research staff can give talks and farmers should be encouraged to voice both their positive and negative experiences of new adaptations.

All agricultural crops require some degree of research into issues such as variety behaviour, cultivation practices and rates and methods of pesticide and fertiliser application. If an operation is large enough it can sustain its own research programme. Smaller companies and individual developers must generally rely on either government services or other industry research. In practice, most projects have a limited research agenda, which is mainly focused on seed collection, demonstration plots and teaching.

Relationship between Farmer and Management

As stressed elsewhere, the development and maintenance of a positive relationship between management and farmers is crucial for the stability of any venture. Adequate incentives, prompt payments, efficient extension services, the provision of timely logistical support and good communication links between management and farmers all play a central role in this process.

The establishment of forums that promote farmer-management dialogue on matters such as contract specifications, agronomic requirements and rectification of misunderstandings and conflict is essential. While almost all contract structures are of necessity hierarchical in nature, farmer participation is essential. Forums offer a channel for experienced farmers to contribute to the structure of contracts and offer advice on local conditions. Managers may also have to address social issues and identify how they can positively contribute to the social and cultural life of the community. This approach cannot be seen as an alternative to effective supervision but as a complementary measure. The three most important issues managers must consider are:

— farmer-management forums;

— male-female relationships; and

— participation in community affairs.

Farmer-management Forums

Intermediary bodies that link management and farmers for purposes of negotiation and interaction are necessary for all contracts. The absence of communication between farmers and managers can result in misconceptions, misunderstandings and, ultimately, confrontation and conflict. By creating farmer-management forums or similar channels, sponsors can negotiate contracts with farmers either directly or through their representatives.

Organisations that are set up to represent farmers need to be truly representative of the farmers and not simply an extension of management with a few token farmer representatives. For example, problems were experienced in Kenya's sugar industry because the organisation supposedly representing the growers, for which the farmers had to pay a levy to cover operational costs, was managed by a board containing only four farmer representatives, whilst there were three representatives from the Government, one from the company and one banker.

Nomination by farmers of at least one representative from each locality provides communication directly between management and the contracted farmers rather than only through field extension staff. Ideally, the representatives should meet with management at least three times in a season. The first meeting should be at the beginning of each season in order to ratify the pricing structure and to discuss the season's crop schedules and any new adaptations that management may want. A second meeting is advisable immediately before harvesting to discuss the crop's progress and to confirm buying procedures. A final meeting to review performance at the end of the harvest may coincide with the final payment to farmers.

In Thailand, a contractual arrangement involving a foreign sponsor highlighted the inexperience of the sponsor. Although the idea of a written contract was new to the area, the sponsor gave little consideration to its details until only a short time before the farmers transplanted their first crops. As a result, the crop was half grown before the contract was distributed. Furthermore, the contract needed four revisions and approval from the sponsor's head office. One unsatisfactory outcome was that the terms of the contract were worded such that farmers' representatives were responsible for the enforcement of the conditions, not the company. In another project in Kenya, farmers apparently did not comprehend or understand the conditions of the contract and expected payments despite the fact that yields did not cover the cost of production support provided.

Both such experiences justify farmer-management dialogue. Furthermore, it is desirable that all farmers attend formal registration meetings at the beginning of each season. Such meetings provide an opportunity to explain management's crop programme, the specifications of the contract and the farmers' obligations under the terms of the contract. Frank public discussions on the formulas of contracts and the clarification of technical specifications are as important as the formal ratification of the agreement itself.

Male-female Relationships

Contract farming can be a catalyst for antagonism between men and women, and this can affect both productivity and farmer morale. Contracts in many developing countries are automatically made with male family heads. In China, a large venture with many thousands of farmers had contracts exclusively in the name of the senior male of each family although, in practice, females did the bulk of the work. When the contract nominee does not do the work the actual workers may not receive a just reward for their efforts. In one project there was a steady decline in crop production and a high ratio of voluntary dropouts in a particular district. On investigation it was found that although wives, sisters and daughters carried out the majority of all crop activities, they failed to receive adequate payment from the registered male contractors. In subsequent seasons the females refused to work, a situation that resulted in the abandonment of many contracts. The situation was corrected by a change of policy by the sponsor, who stipulated that the contract be registered in the names of the actual workers. An incensed reaction by the former male contract holders proved short-lived, productivity was restored and farmer-company relationships improved substantially. Eventually several female farmers were elected as grower representatives to the project's forum.

In Kenya, the Tea Development Authority experienced tensions between management and women. The women were not encouraged to establish themselves as independent tea growers, with the Authority's policy being to favour only male household heads. This resulted in the alienation of active working females. Such examples reflect poor foresight by sponsors concerning the aspirations of female farmers, as well as a failure to consider the subsistence crop requirements of the contracted farmers and their families. In short, attention should be given to allocating contracts and making payments to the principal workers rather than the household heads. It should be recognised, however, that this might be difficult to implement where traditional customs predominate. The critical factor is not so much who is the registered contractor but how the profits are distributed in relation to contribution and work effort.

Conflict is not restricted only to the use of labour and the distribution of economic returns. Land use and different priorities in relation to subsistence food crops and contracted crops can also be sources of intense household disharmony. In East Africa, the introduction of contracted tobacco conflicted with the cultivation of millet, an important subsistence crop. In another example, contracted rice production clashed with the cultivation of sorghum, a crop traditionally grown by women. An impasse was overcome only after the women successfully negotiated a compromise with their husbands.

Participation in Community Affairs

The educational, sporting and cultural activities of the farming community are often very important. The involvement of sponsors and their staff in the local community helps to create a positive atmosphere of partnership. Misunderstandings between sponsors and farmers can be at times rectified at social occasions, a natural follow-on from the more formal farmer-management forums.

A privately owned sugar project in Zimbabwe provided sewerage facilities, piped drinking water, a road network, medical and sports facilities, housing and schools to its farmers. In another venture, charity horse races were jointly organised by management, extension staff and farmer volunteers. Considerable sums were raised and distributed to schools, libraries and health clinics, primarily those within the project's area of operation.

Company policies, based on recognised social and ecological responsibilities, create a positive social environment. Extra-contractual contributions by sponsors to the rural communities in which they operate are not only a positive contribution to the general community but assist in strengthening the system. However, such support must clearly be confined within the boundaries of economic logic, and farming communities should not become dependent on such contributions.

4

Monitoring Performance

All farming ventures need reliable records to provide monitoring, evaluation and diagnostic functions of the key activities that determine the product's performance. Managers need to accumulate, analyse and distribute information not only for the maintenance of project and farmer records but also to evaluate production performance and calculate yield indicators. Such monitoring is particularly necessary in large ventures that involve thousands of farmers, where it is difficult for management to have close interaction with individual farmers.

The basic elements of monitoring, or "process control", are essentially the same in all industrial or agricultural activities: management should identify each component, measure its performance, compare the results to a benchmark and, if necessary, take action to rectify any problem at an early stage.

Monitoring Quality and Yields

A routine analysis should be carried out to ensure that current and future production remains within the quality and quantity parameters required. Deterioration of quality can have serious and far-reaching consequences for any business venture. Quality controls are especially critical for high-value crops such as exotic fresh fruits and many varieties of vegetables. Quantity shortfalls can reduce processing efficiency and jeopardise markets. How a project's extension staff can monitor quality and quantity is discussed below.

Quality Controls

Each venture must develop quality control and monitoring systems suitable for its particular operation. Management must prioritise monitoring procedures and decide how often they should be carried out, in what locations and who should be inspected and assessed. Checking product quality can take place before, during and immediately after harvesting as well as at the time farmers grade their own production and when the products reach the company's processing or packaging facilities.

Quality controls may start as specifications in a written contract or as verbal explanations of quality standards given in both pre-season and pre-harvest farmer-management meetings. Perhaps the most practical method is for management to demonstrate by visual presentation the quality criteria that must be met, either for the entire production or, where appropriate, for different grades. When extension staff at the farmgate or buyers at processing stations purchase crops they should explain to the farmers the reasons why they are grading the crop as they are. Where farmers have mixed their grades they should, if practical, be given the opportunity to regrade.

Some contracts contain clauses which specify that sponsors may carry out certain crop production activities if the farmer fails to meet cultivation specifications.

Under such clauses, management may hire outside labour to eradicate weeds, apply insecticides or harvest mature crops. Such practices can be a frustrating, but necessary, feature of contract farming as the failure of a few farmers to follow recommended practices could jeopardise the production of neighbouring farmers. Some of the major causes of poor quality are the failure to apply fertiliser, ineffective weed and insect controls, disease, immature harvesting and indiscriminate grading and packaging. Quality can also be affected during any on-farm processing and after the raw material has been purchased from the farmer.

In one case involving quality manipulation, inferior crops from non-contractual sources were being sold to take advantage of the higher prices offered under the contract. At the same time, prime quality produce was being siphoned off to the open market for even higher prices. The managers admitted that they had lost control of the buying operation. Infiltration of non-contractual produce from unregulated sources may cause a deterioration of quality standards, the introduction of disease and problems of excessive supply. To stop such practices, clauses in contracts could stipulate the designated origin of the contracted crop. However, this may be difficult to apply in the case of the informal individual developer who is buying crops from farmers over a wide area where it is virtually impossible to enforce agreements. In Uganda, cotton ginners faced with this problem established the Uganda Ginners and Exporters Association, which placed monitors at each ginnery in an attempt to control extra-contractual sales that would lower quality standards.

Where problems regarding quality are encountered, a detailed investigation is necessary to discover why. One such study was carried out to identify why farmers harvested tobacco in a very immature state; a practice that resulted in the serious deterioration of leaf quality. The study highlighted a number of constraints cited by farmers and management that they alleged influenced immature harvesting.

Yield Estimations

Yield forecasting has become increasingly important in agriculture for processing, marketing and budgeting as well as for monitoring of production. Where production is limited to a target quantity, any major shortfall can cause serious embarrassment for the sponsor as a result of inability to meet forward orders. Overproduction will result in expenditure above budgets, unnecessary and expensive stockpiles and quota reductions in following seasons. This will result in income loss for many farmers. The aim of all contract farming ventures is to maximise economic returns. It should be borne in mind that with some crops the highest yield does not necessarily mean the highest income as high yields may only be achieved at the expense of quality or through the uneconomic use of inputs.

There are two basic ways in which yields can be estimated: visual observation by experienced extension workers and, less commonly, by statistical analysis. Calculated yield indicators (CYI) can then identify farmers, villages or districts that may have infiltrated crops in to, or out of, the buying system. Excessive production could occur when farmers overplant their allocated quotas. Extension workers need to verify that the exact quantity allocated has been planted and also need to estimate the potential yield of each farmer. Normally, a farmer's crop yield should be estimated three times during a single season. The first estimate is based on the quota given at the beginning of the season and is referred to as the target estimate; the second is carried out a short time after sowing or transplanting, once the crop is well established.

A third estimate can be made when the crop is fully developed, normally immediately before the first harvest. If marked disparities occur between the final estimate and the actual quantity supplied, the variations must be explained and accounted for. In established projects individual crop estimates per unit area are often based on the mean of the previous three seasons. Yield estimates should be based on knowledge of production in the area or by trial plots that the sponsor has planted before quotas have been distributed to the farmers. If a new crop is introduced to a district, estimates should be based on the productivity of that particular crop grown in similar environments.

Calculated Yield Indicators

A number of methods can be used for forecasting crop yields. These include a top-down model using a hierarchy of variables to the point where optimum performance cannot be improved further, as well as the Baysian model using categorical variables based on experience. Another model for crop forecasting is through the Monte Carlo simulation method based on soil and management parameters. In many cases, however, visual methods based on the knowledge of experienced field staff should suffice to estimate yields.

Higher than expected production justified an investigation to identify those villages introducing produce from outside the contract. The wide variation between actual and calculated yields discovered in this example reinforced the belief that extra-contractual production had been infiltrated into the project. Indicator models provide information for management to identify farmers who circumvent buying procedures. The modelling of a calculated yield indicator is, however, dependent on reliable data and the ability of the assessors to interpret it.

Production Matrices

Simulated modelling of crop production on a scientific basis is a highly complex, ever-expanding and normally efficient discipline. Analytical decision models, such as a production-marketing-consumption (PMC) matrix, are useful in agricultural projects to evaluate managerial, agronomic and infrastructural factors necessary for the success of the crop. The model was first introduced in California in 1979 to structure the production and marketing requirements of a new commercial crop, jojoba. The basic concept of the production matrix, a subsystem of the total PMC matrix, is to formally identify all the key components that influence production in terms of their significance, responsibility and performance. The information on which to base the performance assessment can be extracted from production statistics, farmer records and field observations.

In addition to the farmers, the participants included a multinational corporation, a parastatal marketing organisation and the government's agricultural department, which was responsible for quarantine controls and postharvest pesticide treatment.

Critical (C) components had to be achieved in their entirety otherwise the project would be placed in considerable jeopardy. Important (I) components generally had to be achieved to the maximum extent possible, although marginal performance of one or two such components would not necessarily endanger the project. Significant (S) components were considered important but not vital for the project's success.

Following the first season of fruit production, qualified personnel completed the matrix by rating the achievement. The achievement rankings of the various components showed negative performances on the part of the farmers, the multinational corporation, the marketing authority and the government agency responsible for quarantine, for four critical components. The failure of the harvesting and grading activities, irregularities in the quarantine procedures and air cargo limitations eventually contributed to the collapse of the venture.

Contract farming, which normally incorporates new agronomic and management methods, needs constant feedback regarding the acceptance of new techniques by farmers, disease tolerance of new varieties and the changing work patterns and production capabilities of farmers. A production matrix can be used to provide a blueprint of all factors that should be described, monitored and evaluated. Where possible, such matrices should be developed in full consultation with farmers, for example at farmer-management forums.

Monitoring Human Resources

Managing, motivating and monitoring human resources involves the recruiting, training and deployment of employees as well as an appraisal of their work. The production performance of each farmer also needs to be recorded and assessed so as to determine the need for quota changes and, on occasion, the cancellation of contracts. To support the decisions that managers must make, which can affect the livelihoods of thousands of individuals, there must be detailed monitoring procedures. The two human resource areas that managers of ventures are directly responsible for are:

— appraising employees; and
— reviewing farmer performance.

Appraising Employees

Employees who have been successful in an agribusiness context that does not involve contract farming may not easily make the transition to a contract farming environment. This can apply to senior management as well as to junior staff. Some highly competent and qualified staff members have proved unsatisfactory when required to work with farmers. In projects with only a small number of employees, staff appraisals are relatively simple as the interactions between managers and their staff members are almost on a daily basis.

When a project has hundreds of employees, personnel reviews are generally carried out at the departmental or sectional level. Extension staff can be evaluated through visits to farmers' fields. This should be done at least twice a year. The most practical times for such visits are usually before the first cultivation, a short time after transplanting and at the height of the harvest. This inspection gives the manager and the employee an opportunity to discuss on an informal basis the progress of the crop and the performance of the farmers under the guidance of that employee. The manager can observe at first hand the extension officer's relationships with the farmers, gain an insight of the employee's knowledge of the state of the crop and his or her ability to organise its logistics. The second method of appraisal is to formally review all employees at the end of each season. Their performance can be measured against their job responsibilities and production targets that were estimated at the beginning of the season.

Reviewing Farmer Performance

Formal monitoring of the crop at all stages of its development may lead to the identification of factors that could cause poor production. For instance, the failure of farmers to notice the early signs of disease may delay the application of fungicides, thus endangering both quality and yield. Extension staff must therefore carry out routine inspections of all the farms they supervise. For some crops, such as sugar cane and fruit, inspections may be on a weekly or fortnightly basis. For vegetable crops, several visits a week may be required. Simple report cards should be kept indicating each farmer's code, locality, quota, land unit, the estimate of production and the inputs advanced. The extension officers must document the dates of sowing, transplanting, irrigation, harvesting and any other major activity carried out, such as fertiliser application and chemical use. Comments on climatic conditions, standards achieved and advice given to the farmer should also be recorded. Wherever possible, farmers should be encouraged to maintain their own records of their use of farm inputs and of their production.

At the end of every season the final production, product quality and net income of the farmers can be included in the records. By reviewing such information, management can estimate the farmers' attainable quotas for the following season. Farmer-management forums may then be used to advise future quotas and, if necessary, provide explanations for any modifications to the quota as a whole or for individuals.

Protecting the Environment

Ecological considerations combined with sound agricultural practices should have an important role in contract farming, although the sustainability of the farming system as it relates to the physical and social environment usually receives little attention from either agribusiness or governments. Environmental issues can vary not only from country to country but also from district to district and farm to farm. Deforestation, the depletion of water resources and soil degradation are major concerns that accompany any agricultural development. However, agribusiness is frequently interested only in a single crop whereas farmers usually have permanent multicropping systems and must take the sustainability of these into account when assessing the value of contract farming.

Environmental degeneration as a result of any form of farming can become a major problem if it is not controlled. The cultivation of crops on thin soils on very steep terraces encourages high levels of erosion. Twenty-five percent of crops in one contract farming venture in China were grown on land totally unsuitable for intense cultivation. All participants of the contract - sponsors, the local managers and the village committees - allowed production on steep slopes without any regulation or apparent concern.

A recent survey in Fiji confirmed that at least 30 percent of that country's contracted sugar-cane crop is grown on unsuitable land. Pressures on agricultural land are forcing sugar-cane farmers on leased land to cultivate steeper slopes. These practices were given tacit approval by the parastatal sponsor, despite land-use legislation and evidence that the soils were eroding at a rate two to three times greater than the rate of replenishment. This situation echoes similar attitudes to those found in Tasmania, Australia, where farmers cultivating crops under contract were unwilling to acknowledge that they had serious land degradation problems.

Laws on control of land use are common in most countries but they are rarely enforced. Legislation governing the environment of one developing country includes "at least 25 Acts" relating to ecological policies and there are at least fourteen government and parastatal bodies that administer environmental issues, but the impact has been negligible. Action by sponsors and managers to address environmental concerns is ethically and economically imperative. The most practical way managers can control ecological compatibility is to ensure that all contracted fields are selected by the extension staff in consultation with the farmers. Choice of land should be based on criteria related to soil depth and quality, land slope and water resources. In addition, an understanding and knowledge of previous land usage is important and, if necessary, a crop rotation regime may have to be agreed with farmers. Decisions by management on land use on behalf of the contracted farmers are, however, common, with the extension staff selecting all fields. Experience indicates that when choosing land the opinions of farmers and those of management do not always coincide.

By gaining an understanding of the area and its farmers, managers can assess each farmer's production capacity and the effect the crop may have on the environment. The willingness of the managers to learn from local experiences, sometimes developed over several generations, is important when supervising farmers under contract. In the South Pacific, local knowledge was ignored by expatriate agronomists.

When a government agency introduced bananas for migrant smallholder farmers to produce for export, the customary landowners informed the agency that it could lease as much land as it wanted but "it will never grow bananas". Although the venture was highly organised and adequately funded, the inherent infertility of the soils, a fact well understood by the landowners, resulted in the eventual collapse of the project. Political pressure to settle landless farmers from overpopulated areas, coupled with inappropriate technology, had caused a loss of farmer morale and land misuse.

5

Impact of Contract Farming on Society

Contract farming has been practised for some time now in parts of India and there have been a few recent studies of its impacts on contracting households' incomes. However, these studies rarely examine the impact of the contract farming system on labour conditions and labour relations in India. There have been many studies of conditions for women workers and gender discrimination in export-oriented industries across the world during the past 30 years. While some see the employment of women in such enterprises as a natural economic process driven by business dynamics and labour family needs, others view it in more gendered frameworks and have been critical of the way female labour is treated and paid in these factories.

The whole process of industrialisation is not only export-led, but also female-led. However, female labour conditions in the farm sector under agribusiness expansion have not yet been subjected to similar analysis. Agriculture is becoming increasingly 'feminised' globally as men move out of the sector more quickly than women. The reasons for increasing female representation in agriculture vary from voluntary and involuntary withdrawal of men from farming, to growing wage opportunities for women in export-oriented non-traditional crops and activities.

While these new labour arrangements have led to marginal increases in real income for some women workers, they have also changed relationships between workers and employers, workers and work, and led to differentiation within labour. For example, women's wages are lower and stickier (i.e. do not rise as fast) than men's, working conditions poorer and bargaining power more limited. These are, to some extent, the result of structural factors. Others argue that labour markets are not only bearers of gender, they also reinforce gender inequality.

Women have increasingly been employed to do more delicate work because of their supposedly feminine traits, such as docility, obedience and nimble fingers. But these are not natural or innate skills; rather they are social skills embedded during girls' upbringing as part of their socially defined subordinate position within society. In fact, the expanding employment of women in agribusiness has many parallels with female employment in industry, where women are also preferred for their so-called nimble fingers, and their work is insecure, badly paid, and flexible. In agribusiness, some of these aspects are the result of the nature of production in a sector characterised by seasonality, mono-cultivation and natural risk.

Though there have been studies of gender impacts of contract farming within producer households, such studies of labourers' households are more unusual. In many cases, rural women working in processing plants and related activities reported that employment had given them greater self-esteem and power within the household. However, contract farming does lead to gender inequalities in the quantity and quality of work, as well as wages for women and children, who work longer hours as they are considered better workers, but are paid less. Employment on contract farms is also temporary and sporadic. Some activities, such as pruning, spraying, thinning, pinching out, tying and harvesting of grapes, are reserved for women, and this work differentiation extends even to the processing sector. Such 'gendering' of tasks is a problem as it can exclude women from activities that are better paid, less strenuous or less hazardous.

Child Labour

Child labour is one of the major problems in contract farming, and this is the case throughout the developing world. However, no data on its global or regional magnitude are available because contract farming is practised in small pockets in each developing country. Similarly, there are no estimates of the magnitude of child labour in contract farming in India. The International Labour Organisation (ILO) estimated there were 250 million working children in the world in the age group of 5-14 years in 1996. The work participation rate of children between the ages of 10 and 14 years was as high as 14%, and in some parts of the world, such as central Africa, it was more than 32%.

India is one of the main users of child labour in the Asian region. The 1991 Census of India reported that there were 12.6 million full time and 10.4 million marginal child workers in India, with Andhra Pradesh having the highest number of child workers. In the 1990s, about half the child labour in rural India was concentrated in the five states of Andhra Pradesh, Karnataka, Maharashtra, Madhya Pradesh and Tamilnadu.

Any kind of work by children under the age of 18 is often considered exploitation in developed countries. In India, however, the term 'child labour' applies only to children aged between 5 and 14. The Child Labour (Prohibition and Regulation) Act, 1986 states that no child below the age of 14 shall be employed to work in any factory or mine or in any other hazardous employment. The disaggregation of child labour into different occupational categories reveals that almost 80% of working children are in the agricultural sector.

Since child labour tends to be concentrated in agriculture, it may be reasonable to expect that the incidence of child labour will be higher in agriculturally less developed regions. But, that is not the case; Bihar state, usually referred to as an underdeveloped state, has one of the lowest child labour ratios in the country. On the other hand, Tamilnadu and Andhra Pradesh, more advanced agriculturally, have the highest incidences of child labour. Also, there is a significant correlation between child labour and economic participation rate, especially female participation rate. The states with the highest incidence of child labour are also the ones with a high per capita income and a much higher female work participation rate than many other states. This is because in contexts where there are extended labour opportunities yet a low degree of labour empowerment, poor families are tapped for additional labour power. Since adults usually have some form of alternative employment in such regions, child labour is more likely to be used.

A slight gender bias favouring employment of young girls can be seen across all states. In 1999-2000, the work force participation rate for rural female workers in the age group of 10-14 years was higher than for rural male workers in the same age group. But the overall participation rate of rural females was much lower than for rural males. Employers/contractors prefer girls for a number of labour supply traits like docility, obedience and stability; a phenomenon which is well documented in the literature. Most of the girl labourers belong to the scheduled caste and scheduled tribe households, poor backward castes and other such communities.

The lower status assigned to girls in Indian society, where male preference rules, further propels their employment in a number of activities. Even if parents do not wish to exploit their children, they are often trapped in a vicious cycle of indebtedness, bondage and other obligatory relations which compel them to send their daughters to work. Yet the employers or contractors of girl labour have no requirement to take care of these children; if any health problems arise, they are simply replaced with a new group. With no social security obligations, there is hardly any cost involved for the employers or contractors. Child labour under contracting is not subject to any legal or public disapproval.

Case Studies from Andhra Pradesh and Punjab

The states of Andhra Pradesh and Punjab have been pioneers in contract seed production and contract vegetable production respectively in India. These case studies bring out the gender and child labour dimensions in the practice of farm labour under the contract farming system. The case studies from Andhra Pradesh, mainly drawn from Ramamurthy and Venkateshwarlu and Corta, are based on intensive field research by these authors on the labour processes, conditions and households involved in hybrid cottonseed production. The Punjab case study is mainly based on my own interactions and observations in the field as part of a larger study of contract farming in the state.

Contract Cottonseed Production in Andhra Pradesh

In 1999-2000 Andhra Pradesh had the highest incidence of child labour in India, with 25% of children aged between 10 and 14 in the rural areas working, compared with only 9% in India as a whole. This was also accompanied by the highest rural female work participation rate in the state (48%), compared with only 30% in India in 1999-2000. Furthermore, 47% of the total rural employment in Andhra Pradesh in 1993-94 was casual compared with 36% for rural India as a whole. The proportion of casual employment was much higher for female labour (53%) than male labour (43%). Contract farm labour is generally casual labour, though workers may be tied by advance loans.

This sets the context for the first case study: labour in contract farming of hybrid cottonseed. Three districts of Andhra Pradesh account for 70% of India's cottonseed production and 97% of the total area (28,000 acres in all) under cottonseed production in the state. About 60% of the state's seed production is destined for sale in other states of India or abroad.

Contract farming for hybrid cottonseed production is practised by the MNCs, as well as national and local seed companies, who arrange contracts for seed production with local farmers either directly or through middlemen called 'seed organisers'. These companies or seed organisers supply seed to the farmers, set the price of seed before harvest, and supervise the cultivation of the crop. They also advance capital to the farmers, thus linking markets for output and credit. Hybrid cottonseed production is highly labour and capital intensive. It requires four times more capital and nine times more labour per acre than ordinary cotton production. Labour accounts for 60% of the total cost of cottonseed production, and the most labour intensive operation is cross-pollination, which has to be done manually. It involves about 90% of the total labour input and 45% of the total capital input.

Most of the cross-pollination work is done by young girls who work daily from July to February. Cross-pollination and harvesting (picking cotton) are carried out for six months and four months each by the girls, with some overlap between the two tasks during November to February. The predominant task is cross-pollination, which takes 95% of their time. The involvement of young girls in cottonseed production is so high that it is estimated that 0.25 million girls are employed in this activity throughout Andhra Pradesh. In one case, on a 70-acre cottonseed farm dispersed over 14 different villages, 560 girls were employed and were monitored by 14 supervisors. The girls and their parents are contacted well in advance of the crop season and credit is extended to the parents who agree to assign their daughters to the cottonseed farmers for the season (July-February). This credit (cash/grain) advance, which ranged from Rs. 100-3000 in 1997-98 and was obtained by more than 90% of the girls' parents, serves to interlink the credit and child labour markets; this has become an important part of the MNCs' contract farming strategy. Generally, 10-15 children are hired for 100-150 days per acre of cottonseed production.

Children as young as six years old work from 8.30 am to 6-7 pm; some boys even return to the field again from

9-12 at night. The cottonseed production calendar has been standardised by companies for seed certification and marketing. This has resulted in the regimentation of children's work schedules, who are continuously employed for six to nine months a year. Children's labour is extended well beyond the standard workday (9-10 hours a day with a one hour lunch break). They are paid a piece rate of Rs. 10/- per 25 kg cotton for delinting, and Rs. 0.05 per pest/worm caught. Many girls also accompany their mothers for casual work. They are paid daily wages or piece rates.

Girls are preferred in cottonseed production because their wages are lower than adults', they work longer hours and more intensively, and are generally easier to control. It is reported that one girl can do the work of three adults. Though the agreement obliges these female children to work for only one season, in practice they tend to work for several years for the same farmer. Children are given incentives in the form of chocolates, biscuits or snacks, or prizes and bonuses like a stainless steel tiffin carrier every two years for better and faster work. They are even taken to a cinema twice a month and allowed to watch television or videos while separating cotton from seeds. In some cases, girls are brought in from outside the local area and made to stay in the employer's house and cattlesheds throughout the season.

Whilst young girls are the most popular labour type on cottonseed farms, women are also employed, though for much less of the time and for fewer tasks. The most common labour arrangement for women in contract farms is daily casual work. Though this work ensures instant wage payment, in practice the wages are held back till a task is completed. There is also increased micro management of labour time and supervision of daily wage labour by landowners, who may threaten women workers with losing their jobs or withholding their wages. Labour time is controlled by the landowners in various ways, such as going to workers' homes each morning to fetch them early for work, ensuring that lunch is brief, prolonging the work day, and insisting on workers making up for any tardiness in work.

Often, even the supervisors are women. In order to reduce the need for supervision and control, another labour arrangement has emerged, especially for cotton picking and weeding. The piece rate system leads to workers lengthening the workday themselves, magnifying the pace and intensity of work and even monitoring each other's performance. Daily wage labourers pick about 10-15 kg of cotton a day; those working under the piece rate system can pick as much as 20-30 kg, with some even reaching 40 kg. This leads to a change in the relationship between efficient and less efficient workers and creates a hierarchy of workers. As individual wage earnings become more uneven, average wage costs for producers are reduced.

Since the 1980s the cross-pollination work in cottonseed farms has shifted from daily adult labour to almost entirely labour by girls, who are fully bonded i.e. obliged to work only for the contract grower who advanced the loan. In fact, adult women are discouraged from cross-pollination work, and girls over 14 or those who have reached puberty at the time of agreement are not permitted to work in contract farms by the seed producers, organisers or companies. Many rumours are spread to discourage women from undertaking such work, such as that post-pubescent girls working in the fields will cause crop failure. This is no different to countries like Japan and Korea, where until the 1980s married women were prevented from working by law or strong social customs.

The employment of female children in cottonseed farms has had many impacts on gender relations at the household level. For example, girls have more responsibility for household provisioning, and with more money coming in, men may withdraw from work and often resort to drinking. There are also health implications for girls involved in cottonseed work; higher levels of pesticides are used in this crop than in ordinary cotton cultivation. Girls working in cotton fields for many years have been reported to suffer menstrual problems. Further, more and more girls are being taken out of school to do such work.

Contract Tomato and Potato Production in Punjab

Agricultural labour accounts for as much as 88% of the total rural labour in the state of Punjab. More than two-thirds of the agricultural labour in the state, including contract farm labour, is casual. Major reasons behind the casualisation of farm labour in the state include mechanisation of major farm operations, inflow of migrant labour, slowdown of agricultural growth, and non-viability of small and marginal holdings.

Women's labour participation rate in agriculture in Punjab has been extremely low, though higher than their participation rates in the national, rural and urban workforce (2.8%, 2.2%, and 4.3% respectively). Only 1% of cultivators (farmers who own or lease land) were female in 1991. Punjab also has had one of the lowest incidences of child labour (2.7%) in the country. However, Punjab has been a pioneer in India in the contract farming of perishable produce, a sector which tends to rely on women and children's labour.

Fruit and vegetables require more intensive labour than other crops. Contract farms under company supervision are little more than 'factories in the fields' from the point of view of labour. Their labour system resembles the industrial sector because of timing, quality and standardisation requirements which cannot be met by mechanical methods. This requires 'quality labour', i.e. efficient, timely and paced; 'flexible labour', that is readily available and cheap; and 'docile labour', in other words politically trouble-free labour. These conditions are met more easily by women, who are perceived to be homemakers and, therefore, low cost, sincere and more obedient workers. This amounts to manipulation of the cultural understanding of gender.

Contract farming of vegetable crops has been practised by both MNCs and local firms since the early 1990s in Punjab. The contracts specify quantity/acreage, quality, timing and price of the produce. The companies also provide seedlings on part credit and specify the production practices and inputs to be used. That contracting has led to more and better employment opportunities for labour, especially women, is true and acknowledged by the labourers. The labour intensity of vegetable crops, except potato, is much higher than for traditional crops like wheat or paddy. The labour intensity is 3,600-4,000 hours per hectare for tomatoes, compared with only 740 hours per hectare for paddy. This has created a big employment boom in the contract production areas of the state, especially as the mechanisation of sowing and harvesting operations of paddy and wheat crops has reduced manual work to almost nothing.

While employment has grown as a consequence of the labour intensive nature of the crops, accompanied by the emergence of some employment in processing, wage levels have been pushed to subsistence levels as a result of the increased competition. This has come about through in-migration and the cessation of out-migration because of the higher incomes available locally. At the same time, those in work have to deal with insecure employment and poor working conditions.

According to Gill, during the mid-1990s, three-quarters of all workers employed in the state's vegetable production sector were hired labour. Further, female labour accounted for 58% of total labour hours, compared with 34% in paddy. And 49% of all those working in the vegetable production sector were hired female labour, compared with 25% in paddy. In tomato production alone, female labour accounted for almost 60% of the total labour hours. Child labour accounted for about 3-4% of the total labour hours in vegetable crops, as part of family labour. However, a woman's wage is only 75% of a male worker's wage; a child worker receives only half that of a male worker.

Labour in Tomato Fields

Women and girls mostly transplant and harvest tomatoes as such workers are more available, cost less, are more honest and better suited to picking and transplanting jobs. Mothers with infants also work in contract farms and infants and children remain on the farm throughout the day, with implications for their health and nutrition.

There is a piece rate system of wages linked to output, i.e. Rs. 2-2.5 per crate of tomatoes picked (a crate contains 20 kg of tomatoes). Thus, the more a woman or girl picks in a day, the better she gets paid. Harvesting one acre of tomatoes takes 15-20 women two days as one woman can pick only 25-30 crates a day.

Wages go up when tomato harvesting competes with that of potato and wheat in some areas. But wages soon become depressed with the large seasonal inflow of migrant male labour to the state. In fact, migrant agricultural labour accounts for 25% of the total agricultural labour force in the state. The daily wage rate for female workers is only two-thirds of that received by men. In tomato harvesting, half-day wages are also common as harvesting should be completed before noon.

Labour in Potato Fields

Potato cultivation, especially under contract, has increased women's employment; a large number of women can be seen picking and grading potatoes during harvest time, though in general, female labour accounts for only 38% of all labour used in this crop. Generally, female labour is preferred for these jobs as they require patience and care, yet women's pay amounts to only two-thirds of that received by men. A lot of child labour is also used, especially for grading. Even mothers with infants can do such work, which is generally in one place and under the shade of some tree or in a shed.

Daily wages are generally paid for harvesting; usually Rs. 40/day for women and Rs. 20/day for girls. Groups of workers may receive a fixed payment contract to harvest an entire potato field; this was Rs. 500 per acre in 1999-2000. Bagging potatoes is a separate labour activity and is paid at the rate of Rs. 5-10/bag. In some cases, women members of the contract grower households supervise potato grading, especially when the produce is graded at the grower's farmhouse. This is preferred as most of the grading labour is female, and it is easy and more effective for a female member of the farming household to control their work.

6

Contract Farming in Developing Countries

New types of agricultural production and marketing can occur under many different types of institutional arrangements. These can include plantations or state farms, nuclear-plasma combinations of plantation and small-holdings, various land and labour sharing arrangements as well as traditional smallholder family farm operations. The most common arrangement in both developed and developing countries is where land-owning farmers sell to local or city spot markets receiving prices that reflect purchasers' valuation of produce on the day based on quality and quantity.

There are no overlapping contracts such as the purchaser providing credit and no forward pricing arrangements. However, there are alternatives to spot markets and these arrangements can be complex. Forward integration can occur where a group of farmers own or control a marketplace or backward integration occurs where large processing and marketing firms either own farms or become directly involved in supporting and controlling production through contracts.

The latter type of arrangement is called 'contract farming' and usually involves a large agribusiness firm integrating backwards by forming alliances with groups of smallholders and, through written or verbal contracts, providing farm inputs such as credit and extension in return for guaranteed delivery of produce of specified quality often at a pre-determined price. Such contracting arrangements may also involve horizontal integration where firms not only provide direct inputs into farm-level decision making but also encourage integration of various activities across a population of smallholders through farm groups. These groups may coordinate planting and harvest as well as facilitate or manage storage and transport arrangements.

There are probably as many types of contracts as there are contracted smallholders. However there are common elements that distinguish 'contract farming' from alternatives such as plantation farming, share farming and selling through local markets. Contract farming impacts on marketing of produce and usually at least one of the other three stages that comprise an agricultural micro-system: input supply, production and processing. The simplest contracts are usually restricted to some type of forward selling.

For example, Mangosteen producers in Bali receive an early season payment in return for assurances they will deliver the harvest to particular exporters who pay them the balance of the prevailing market price at delivery time. Other contracts are more complex. East Javanese seed corn producers contract with a multi-national seed firm using contracts that include quality and quantity clauses, provision of credit and require meetings of the farmers' group with firm extension officers to learn about seed crop management. These smallholders are tightly controlled in their use of fertilisers and pesticides and other management decisions such as planting density and timing of planting and harvest and the types of companion crops that are allowed.

Spot Markets and Plantations

Contract farming is an attempt by usually large agribusiness firms to expand their operations to improve profitability or manage risk by diversifying their sources of agricultural products. To understand why expansion may involve contracting with smallholders rather than, say, by accessing spot markets more aggressively or by developing private plantations, 'transaction costs' encountered by large firms wishing to expand their operations need to be considered. Williamson states 'economic institutions evolve to minimise the costs of resource allocation'. This means that of the plethora of structures a firm might take the firm eventually adopts a structure that minimises the (transaction) costs of assembling the resources necessary to meet demands emanating from markets.

Three factors contribute to transaction costs: bounded rationality, opportunism and asset specificity. In the absence of these factors, contract farming would not occur since agribusiness firms could buy all their produce in spot markets which would be instantly and perfectly responsive to their demands. Bounded rationality describes differences in information between contracting parties. For example, the firm may have an excellent knowledge of markets while the smallholder may have little knowledge of them hence may benefit from a contract.

Opportunism may occur when there are opportunities for taking advantage of situations to the detriment of the other party in an agreement. For example, smallholders may be concerned the firm could, by virtue of its market domination, offer a very low price in the spot market or, alternatively, the firm may worry that sellers could collude to drive up prices. Writing contracts clearly spelling out obligations may reduce these types of concerns. Asset specificity is the third factor contributing to transaction costs. It reflects risks associated with protecting 'sunk costs' in processing plants, logistical systems or market development or, for smallholders, costs of protecting investments in specialised machinery and knowledge. Both firm and smallholders may protect these investments through contracting.

The major alternative for the agribusiness firm seeking to expand its activities is to source from the open market. This is the usual arrangement with staples such as cereals and standard livestock products. However, this arrangement is less likely when HVF products are involved. High transaction costs associated with the open market arise from uncertain quality and supply and price risk. Food processors usually face tight quality requirements that may be difficult to meet in open markets. These requirements reflect technical aspects of preserving, packaging, freezing and transporting HVF products and, also, meeting consumer requirements.

Consumers expect almost perfect product uniformity for food products. This is particularly so in the United States market where labelling ensures quality and consumers view product uniformity as a quality indicator for both eating experience and safety. Open markets such as that for rice are high volume and therefore are unlikely to have uncertain supply and quality issues can be dealt with cheaply through grading. Also, forward and futures markets exist for shifting price risk.

Hence, these markets function well on an open basis. However, High Value Food (HVF) crops are usually produced in much lower volumes and sold to a relatively small number of purchasers who possess processing capacity to meet demands from consumers. This type of market structure, with a high concentration of intermediate buyers, is not conducive to open market selling since farmers can be 'at the mercy' of a few opportunistic agribusiness firms. If farmers avoid these types of crops, agribusiness firms may find spot markets thinly traded and unreliable as sources of supply. Contracts with individual smallholders are needed to encourage them to undertake production. Finally, thinly traded markets that have periodic shortages usually also have volatile prices. The latter can be costly for processors if they are servicing large downstream contracts written in advance of a season.

Another procurement option for agribusiness firms is for firms to operate their own plantations. This type of sourcing also incurs a number of transaction costs. One of the most important is the cost of supervision which, because of the likelihood of 'shirking', can be high for some types of crops or livestock enterprises. Shirking problems, also referred to as 'moral hazard' or 'principal agency' problems, reflect the high costs of monitoring labour effort when there is uncertainty about yield. Uncertainty over crop yield means the firm cannot distinguish between production lost through lack of work effort and production lost to weather and uncontrollable biological factors.

In principle, even a very slack work team may, in a good season, produce a crop with a good yield. This type of transaction cost is related to the type of crop technology used and increases with its complexity. For example, large plantations work well for tropical beverage crops where management tasks are clearly defined, usually uniform and do not require judgement or much initiative on the part of workers. Alternatively, HVF crops, such as vegetables for export, are unlikely to be successful in a plantation environment since they are technically more sophisticated and require worker initiative to achieve satisfactory yields and meet quality requirements.

Other high transaction costs in plantation production include costs associated with land and skill acquisition. Land acquisition costs may be reduced through leasing or through subsidised land acquisition from government land reform policies. However, even under these circumstances, sunk costs are incurred in establishment of infrastructure and 'setting up'. Plantations also depend on a relatively skilled supervisory and management team requiring substantial investment in human capital. These workers are employed on a permanent basis and are costly. Transaction costs on plantations may also result from unionisation or politicisation of workers resulting in opportunistic behaviour. Also, governments or individuals with political power may see the relatively illiquid investments associated with plantations as opportunities for either ad hoc taxation or as soft targets for extortion.

Despite potentially high transaction costs from plantations, an agribusiness firm may face some distinct advantages from a vertically integrated operation involving firm-owned plantation production and processing and marketing operations. Supply may be more predictable, which is important if the agribusiness firm needs predictable throughput to achieve scale economies in processing or has forward obligations to service large downstream contracts. Also, while opportunistic behaviour by plantation workers may be costly, other types of opportunistic behaviour that occur in out-sourced or contracted production may be avoided.

For example, costs of contract default where credit or other inputs provided to contracted smallholders are lost due to diversion of production to other buyers can be avoided. Costs of resolving conflict may be lower for plantations than for contract farming. Management can use its power to resolve disputes and, because information sharing occurs more easily in integrated firm structures, issues arising from differences in information held are less likely to occur.

Transaction Costs

Contract farming may have high transaction costs however still represent the 'best bet' for expansion of the firm when compared to alternatives. Transaction costs incurred by agribusiness firms engaged in contract farming fall broadly into four areas:

Costs of Drafting Contracts

The initial costs of drafting contracts are increased by the necessity of having first season contracts work well. Glover and Kusterer report that farm contracts have a 'honeymoon' period in their first season where smallholders show high levels of goodwill towards the contracting firm. However, smallholders' experience of the first season flows into later seasons when, with contract maturity, contract provisions tighten and smallholder attitudes harden as a more business-like attitude develops. Hence, getting the contract 'right' in the first season should be a priority for the agribusiness firm and likely to be costly. Upfront costs in negotiating and managing contracts include:

— Costs of gathering agricultural, social and economic information about an area or region.

— Costs of contacting and establishing relationships with individual smallholders, farm groups and village committees or headmen. This may involve both political action and purchase of goodwill at the community level.

— Costs of gathering information on individuals to select smallholders suitable, from the firm's perspective, as partners in contracts.

— Costs of negotiating with individuals including possibly farm and family visits and establishment of personal relationships between smallholders and firm representatives.

— Costs of writing contracts and, where literacy is limited, legitimisation of contracts through a village committee or headman.

— Costs of enforcement of contracts.

Maladaption costs

A pervasive theme in the literature on contract farming is quality specification of contracted production. Smallholders may face difficulties in meeting quality requirements regarding fruit or vegetable size, colour, blemishes and ripeness or seed grain moisture and purity requirements and, increasingly, food safety requirements related to chemical use. The reasons for smallholders apparently having more difficulty than developed country farmers in meeting quality specifications are not well understood. Glover and Kusterer point to economies to scale in quality management and the relatively small scale of production that most smallholder contractors actually achieve. Key and Runsten argue that quality control for HVF crops requires more capital intensive production than for other crops and hence the problem is essentially one of smallholders not being able to afford appropriate technology. Also, quality problems may arise for cultural reasons. The stringent requirements of consumers in developed country markets may require a production mindset that is simply alien to nth generation traditional smallholders used to servicing markets where branding and product uniformity are not so important.

Set-up and Running Costs

The governance of contracts requires a management team to support personal contact with smallholders in the village or region where the agribusiness firm is operating. Management personnel are likely to be involved in dealing with contractual problems such as payment issues and monitoring activities such as farm visits. This management effort may need to be integrated with extension services offered at either individual smallholder or farm group level or these may be run separately with extension services outsourced by the firm.

Bonding Costs

The agribusiness firm has an interest in 'bonding' the smallholder to the contract by subsidising various types of capital acquisition. As the number of successfully completed contracts increases, smallholders gain specialised knowledge about the contracted crop and may invest in sheds, plant or field equipment to expand production or make it more efficient. This build up of capital should cause a bonding process making contract renewal more likely and leading to a long term relationship between smallholder and firm. Some frozen vegetable contracts in Mexico have operated over many years and apparently reflect very long-term expectations on the part of smallholders.

Smallholder Participation in Contract Farming

Contracts are initiated by those who write them and, as discussed above, agribusiness firms write contracts to either increase profits by expanding operations or reduce profit variability by diversifying supplies. However, if contracts are to work, they must be attractive to potential contractors. Hence, contracts struck with smallholders need to increase farm profits or reduce risk exposure.

Taking on New Enterprises

A farmer considers three factors prior to undertaking a new enterprise or substantially changing the way an existing enterprise is pursued. These are revenue implications, cost implications and any additional exposure to risk that might arise from the new activity. In principle, a contract does not need to improve revenues to be attractive if revenue losses are more than fully offset by cost savings. However, researchers have yet to report examples of contract farming involving expected reductions in revenue and increasing revenue is likely to be important.

Two types of farm costs are relevant to adoption: opportunity costs and input costs. Opportunity costs occur with a new activity because other on-farm or off-farm activities may need to be curtailed. In essence, new enterprises 'crowd out' existing activities resulting in lost income from these sources. Hence opportunity costs reflect how resources such as labour, land, buildings and machinery are used in existing on-farm and off-farm enterprise or in leisure or community activities. Labour may be used to generate income or diverted to leisure or community activities. Land is likely to be used for other cropping or livestock purposes or rented out, sheds may be used for on-farm activities such as grading, processing and storage or for industrial piece-work and machinery can be used either on the farm or rented out.

Two considerations are relevant to evaluating opportunity costs. First, it is important whether the input is constrained in the sense that more of it cannot be purchased or obtained through negotiation with third parties. For example, if land can be rented in then the opportunity cost of land used in a new contractual enterprise equals its rental value. In contrast, if extra land is not available and the smallholder is 'land constrained', then the opportunity cost of land in a new contract equals the gross margin of the existing on-farm activity that is curtailed to accommodate the contract. Smallholders may be constrained with other inputs. Hiring in of additional labour may be impossible because it is unavailable and credit may be unavailable because the smallholder has no collateral.

The second consideration in evaluating opportunity costs concerns farm inputs that have zero or very low value for the smallholder. De Janvry, Fafchamps and Sadoulet extend the theory of 'shadow pricing' to farm input markets where inputs such as land and labour cannot be used because of high transaction costs. An example is where a farm family has surplus labour because it cannot rent in more land to provide employment for family members and things like travel costs, availability of work, social factors or cultural beliefs prevent off-farm work.

In this situation, even though the village day rate for labour may be Rp.100,000 the shadow price for the family's labour is zero or very low reflecting only subjective valuation of leisure time or participation in village activities. In this sense, the family has a reservation price for its labour, say Rp.50,000 per day, that is never met by the market after transaction costs are subtracted and hence family labour is under-utilised. Relatively low shadow prices can occur in more subtle ways with land where social factors prevent certain types of activities. An example of social transaction costs in land use is wealthy land owners in Pakistan producing livestock, never crops. Clearly, if a farm contract can provide employment for resources that have very low shadow prices then it will be attractive to the smallholder.

The second type of cost is expenses incurred in purchasing farm inputs. Such expenses include hired labour, machinery, seed, chemicals, storage, marketing and rented land. A new enterprise must increase total farm income if it is to be a candidate for adoption by the smallholder. Increasing total farm income requires expected revenues to be greater than purchased input costs plus opportunity costs.

Adoption also requires farm risk remain at acceptable levels hence the effect of the new enterprise on stability of income is important. How farmers respond to uncertainty is contentious however elements of the problem are captured in safety-first theory, where smallholders will not expose themselves to the risk of their income falling below some minimum threshold level. Farm bankruptcy is relatively common in developed countries where its consequences are serious though not necessarily dire. In developing countries losing land can be dire and empirical evidence shows developing country farmers generally take more conservative account of uncertainty than developed country farmers.

Causes of Smallholders Enter Contracts

Agribusiness firms are likely to have four areas of

strategic advantage that allow cost savings to be conferred on smallholders through contracting:

— They may have access to product markets where high transactions costs effectively prevent smallholder access.

— They may have access to relatively inexpensive credit where, for a range of reasons, smallholders face high interest rates or have no access.

— They may provide services for managing on-farm risk.

— They may provide information on extension, logistics and marketing at relatively low cost.

Access to markets

Recent expansion of contract farming is often viewed as part of the broader globalisation phenomenon whereby removal of trade restrictions has led to increased flows of agricultural products, especially from developing to developed countries. Runsten documents a range of contracts since 1989 for HVF crops including strawberries, melons and frozen vegetables processed in Mexico then exported to the United States by both domestic and multinational agribusiness firms. Goodman and Watts document the development of contracts, alongside other multinational activity, for pineapples and bananas from Central American countries for export to the United States and Europe. Glover and Kusterer document similar activity in Central American countries and Porter and Phillips-Howard examine a range of new types of contractual arrangements involving international trade from Africa.

Agribusiness firms are instrumental in 'opening' markets for smallholders in all of these studies. These firms have advantages over smallholders in market knowledge and experience, information links, legal expertise, economies to size in processing and transport and have the financial muscle necessary for sustaining international trade relationships. From a smallholder perspective, in the absence of contracts these markets are 'missing' in the sense that transaction costs of accessing them on a small scale are effectively infinity.

Access to credit

Non-traditional or HVF crops are more costly to produce than traditional crops and cash requirements for farm inputs are usually relatively high. This is because HVF crops often require specialty inputs and have more exacting quality requirements requiring sophisticated technology and flexible use of labour and chemicals. Hence, smallholders need access to credit to undertake production. Many smallholders are credit constrained in the sense they have no access to credit at all.

Alternatively, if access is available they face high interest rates, often three to four times the bank rate, from local moneylenders or excessive transaction costs if they use bank credit. High interest rates reflect relatively high costs faced by local moneylenders in sourcing funds and servicing borrowers who do not have collateral. Titles to land may be traditional rather than legal and court processes slow, expensive or ineffective so that small loans are not worth pursuing legally.

In this situation, a moneylender is limited to collateralising the smallholder's desire for future loans and, since they are often part of the same community, may face social pressure to be benevolent when repayments are delayed by bad seasons or exigencies such weddings and funerals. These high costs are passed on to borrowers by moneylenders in the form of high interest rates. When smallholders seek credit from agricultural banks or micro-lenders, transaction costs are high. On even small loans they may face forced purchases, loan delay costs, travel costs, application fees, legal service costs and collateral titling costs.

There are several areas of potential savings for agribusiness firms in providing credit. If the firm is large and well established it is likely to obtain funds at normal business rates. A large firm may also have advantages over moneylenders in management of risk because of the size and diversity of its loan portfolio. That is, investing in a large number of small cash advances allows diversification of lending risk either across participants in a particular contract or across participants in a number of different commodity contracts or activities.

The agribusiness firm also has lending advantages by virtue of its contract. A contract allows monitoring of input use, a degree of control over crop management decisions that might jeopardise repayment and can specify how cash advances are to be repaid. Also, contracts require delivery to the firm hence cash advances can be deducted from post-harvest cash settlements.

Other loan protection devices include making future contracts depend on meeting repayment clauses in current contracts and making loans in the form of specialised agricultural inputs rather than cash. Finally, there may be no other local market for the contracted commodity than the agribusiness firm thus ensuring diversion of collateralised farm output cannot occur. These factors reduce the need for collateral and mean that agribusiness firms are likely to incur lower costs than moneylenders which can be passed on to smallholders through contracting.

Management of risk

Farmers in developed countries have three basic approaches to the management of risk. First, they can diversify over both on-farm and off-farm enterprises offsetting losses in one activity against gains from other activities. Second, they can smooth income over time by adjusting savings and borrowings to offset low and high income years. Third, they can use futures or forward markets or crop insurance policies covering price and yield variability to reduce risk. In contrast, developing country smallholders have limited strategies for managing risk.

Diversification on the farm offers prospects for income smoothing however yield risk may be correlated across enterprises if seasonal conditions are the major determinant of crop yield. There may be scope for scattering plots across a district to diversify yield risk or finding off-farm work or entrepreneurial activity when farm returns are down. Restrictions facing smallholders in credit markets associated with high interest rates and lack of access, discussed above, limit the scope for using borrowings to supplement income in bad years and smoothing income over time.

Finally, sophisticated markets for insurance, such as crop insurance and forward and futures markets, usually do not exist in developing countries and remain the province of wealthy farmers who can access international hedge markets. Hence, given the difficulties smallholders face in shifting risk, it is not surprising risk aversion is central in smallholder decision making and particularly so in adoption of new crops and technologies.

Entering a contract may mitigate or exacerbate smallholder risk. If upfront investment is required then failure of either the crop or the contract results in loss.

Alternatively, if the contract works and becomes integral in the farm plan then it constitutes a form of diversification and may reduce risk providing it does not dominate the farm plan.

Non-traditional crops are likely to be more risky than traditional crops. They have higher production costs hence more income is at risk in the event of crop failure. In addition, prices of non-traditional crops are more volatile due to thinly traded markets, yield is more uncertain than with traditional crops and such crops are often more perishable.

Hence, adoption of these crops can be unattractive from a risk standpoint without some form of risk protection. Such protection can occur in contracts in different ways. Subsidies may be provided when farmers first enter contracts to reduce risks in setup of the new enterprise, cash assistance with operating costs and extension and management input from the firm may reduce yield risk. Glover and Kusterer report that smallholders with contracts were subsidised in the early years of their participation and extension from the contracting firms was important in reducing yield risk.

Agribusiness firms could hedge some price risk for HVF products in options and futures markets to protect their own forward commitments and provide upstream protection to smallholders. However, there is not convincing evidence firms actually do this by putting price guarantees in contracts. We found from interviews with Indonesian smallholders that contracts for seed corn, Mangosteens and ginger stipulated prices based on market prices at time of delivery.

Thus, opportunities for reducing smallholder risk through contracting include diversification into a new crop with price movements largely independent of those for traditional products, reduced risk associated with start-up costs and seasonal operating costs met be the firm through subsidies at start-up and forward payments and reduced yield risk from the firm's extension activities.

Crop Specific information

Information can be expensive to gather and is not depleted by use. Hence an agribusiness firm spreading information over many contracts has advantages in providing crop specific information over smallholders gathering their own information. Most contracts described in Glover and Kusterer included visits by firm extension officers to either individual farmers or farm groups several times during the first year of the contract but often less in later years. These visits combined dissemination of information with suggestions about management as well as providing firms with feedback on issues between themselves and growers.

Most developing countries have government extension services to disseminate information about traditional crops however, given the limited nature of developing country government resources, these agencies are unlikely to provide specialised information about new crops. Such specialised information may concern chemical restrictions related to food safety requirements in specific markets, timing of planting and harvest to meet markets, management of product quality and other market and technical information.

Criteria for Success in Contract Farming

Criteria for success of particular contracts in enhancing welfare can be derived from consideration of how contracts work. If contracts are entered freely and there are no barriers to exit, then persistence of contractual arrangements over time indicates both parties believe they are better off and hence the contract can be said to 'successful'. Thus, success of contracts might be measured by whether they persist over time indicating both partners are satisfied with the arrangement. In Indonesia contracts for seed corn, mangosteens, ginger and fresh vegetables for the tourist trade were not signed lightly by smallholders, who often spent a season making the decision. Of interest, all of these smallholders had some experience in production of the crops (or very similar crops) previously and in some cases with the contracting firm.

Thus, entering the contract seemed to be part of the evolution of their selling arrangements rather than something new. There was no suggestion of pressure from agribusiness firms to enter contracts. We found no barriers to exit except implicitly with seed corn where a smallholder exiting the contract would face village level pressure not to grow non-contract corn while neighbours remained in the contract. The latter reflects a technical constraint in production of hybrid seed where neighbouring corn crops can interfere with pollination.

Defining success in terms of the contract continuing over many seasons may be too narrow. Glover goes one step further and defines success in terms of the (1) continuing viability of the contract and (2) distributional effects of the contract. Contracts have community level impacts and favour some groups over others. The contract can create groups of losers that can be relevant in assessing the success of a contract from a development perspective.

Economic, Technical and Social Environment

Whether a contract succeeds depends on uncontrollable

factors in the legal, social, economic and physical environment:

Strong markets

The market environment is essential to success of a contract. Demands met by the agribusiness firm through contract sourcing need to be both strong and not too volatile if contracts with smallholders are to succeed. Contracts between firms and smallholders have considerable start-up costs and a period of low demand for the final product can destroy continuity of a contract as it matures over a number of seasons leading to contract abandonment and losses. In Indonesia, an export ginger pickling plant sourced through contracts and costing several million dollars ended up 'mothballed' after two seasons with grower contracts not being renewed. This occurred because large-scale entry by Chinese ginger producers to the Japanese pickled ginger market undermined demand for Indonesian pickles.

Macro institutional policies

Developing country governments pursue a range of policies to address strategic concerns, improve incomes and wealth and ensure economic and social stability. These policies operate across society and are designed to meet the needs of the majority of citizens rather than particular sectoral interests such as agriculture or exporting. As such, these policies are necessarily imperfectly targeted and may inadvertently work against particular interests. The macro policies likely to influence contract farming are land ownership rules, taxes, exchange rates and food security.

Landownership rules, particularly those preventing foreigners holding land, may favour contract farming by preventing plantation development by multinational corporations. An additional issue here is sovereign risk that arises not because governments prevent foreign ownership or large holdings but because they might do so in the future.

Many farm contracts supply either export or import-competing markets hence volatile exchange rates can lead to difficulties since revenues are earned in one currency while costs are incurred in another. Thus stable exchange rate regimes favour contracting and unstable regimes place contracts at risk.

Food security policy is supposed to ensure that populations in developing countries are well fed. Such policies often use two instruments to achieve goals: self-sufficiency in production and control over food prices. Self-sufficient production of staples, usually meaning not too much dependence on food imports, is usually achieved through subsidisation of inputs such as fertiliser or other chemicals and, in some instances, by compulsory marketing of staples through government controlled boards empowered to sell at prices differing to those prices received by producers. Food security policies do not necessarily conflict with contract farming since smallholders contracted for HVF or exotic production may benefit from subsidised farm inputs.

However, usually subsidisation of traditional production means producers of non-traditional products face greater competition in markets for inputs and outputs. Such distortions, while perhaps justified in a macro context, work against the interests of agricultural expansion based on contracts for non-traditional crops. Also, policies such as requiring firms to allow intercropping, although possibly laudable from a food security standpoint, may discourage contracting.

Sophisticated technology

Sophisticated technology within the contract environment may encourage contract participation. Contracted crops usually require sophisticated technology. These crops require specialised information to produce, are relatively capital intensive and require high levels of management initiative that may not be possible with hired labour. As discussed above, smallholders have limited access to credit and information hence there are synergies from contracting with agribusiness firms providing capital and specialised extension support.

Land ownership

If land tenure is not clear from a legal perspective then land is not useful as collateral for loans. Contracting firms then have a major strategic advantage in interacting with smallholders and access to credit becomes a reason for contracting. If land tenure is uncertain, say because of land reform policy, smallholders need to be more risk averse and are unlikely to avoid significant sunk investments in either new enterprises or land improvements. Flexibility in contracting may provide a way around problems associated with such uncertainty. Short leases in land settlement schemes or other types of government land reforms creating uncertainty over ownership are also likely to favour contracting with its lower start-up costs.

Management of Environment

Ultimately success of contract farming hinges not only on the economic and policy environment but also on the management environment. In this regard, two groups of issues stand out. The first is quality of management, the second, types of actions taken by management.

Porter and Phillips-Howard describe contract management failure in Africa resulting from use of expatriate staff with inappropriate cultural values in management roles. They argue that lack of knowledge of cultural values contributes to under-performance and failure of contracts through misunderstanding of issues, inappropriate conflict resolution processes and miscommunication. They argue for employment of local staff in contract management positions in contracting firms.

Smallholders are vulnerable to corrupt behaviour by agribusiness firm employees at several points in a contract. Corrupt employees may get kickbacks from purchases of farm inputs supplied to, and eventually paid for by, contracted smallholders. Delivery may also present opportunities for corruption if different quality standards or delivery schedules are applied to different growers or if short-weight occurs.

Other aspects of management concern specific actions taken by management to facilitate achievement of successful contract outcomes. Organisation of farm groups, selection of farmers, management of contract default and conflict resolution are described below.

Farm groups

Farm groups can play an important role in success of a contract by encouraging adoption of new technology and adjustment to changed market conditions and by lobbying to deal with political change. Such groups figure prominently in the contract farming literature where Glover and Glover and Kusterer and others see farm groups facilitating contract management by dealing with disagreements between growers and contractors and assisting transfer of technology through group discussions. Bingen argues that successful farm groups are built on pre-existing groups, are democratic rather than 'top down' and they have either explicit or implicit constitutions and protocols. Little and Watts stress the need for groups to be politically independent and self-financing to be effective.

Farm groups can be particularly effective through 'linkage dependence' where there is linking of the purpose of the group to the activity of the contract. Coulter et al. report that contract linkage dependent groups in Kenya achieved better bargaining positions for members, developed long-term investment strategies and improved crop production skills. Members of farm groups often have diverse interests and Glover argues that this leaves them open to subversion by both agribusiness firms and local governments or both working together. Diversity of interests arise because some growers undertake service activities associated with the contract such as transporting produce to the processor or procurement of inputs. More experienced growers may be sources of extension information for less experienced growers and large growers may provide machinery to small growers. Resentments arising from conflicts of interest may lead to risk aversion amongst group members and unwillingness to take action.

Farm groups are important in managing agricultural contracts however it is not clear where the impetus for such groups should or actually does come from. Farmers may have pre-existing groups that can be linked to the contract or a community might encourage formation of a group in response to lobbying from an agribusiness firm. We found in the Indonesian seed corn case the agribusiness firm built a public hall as a gesture to the community in which its contract was being implemented and the hall was used by the contracting group of farmers for meetings.

Selection of farmers

Agribusiness firms selected smallholders for contracts and the possibility of self-selection by smallholders is referred to only indirectly. This reflects that contracting firms are usually perceived to hold the power in relationships with smallholders and, by implication, can pick and choose partners for contracts. The distinction between selection by the agribusiness firm and self-selection is important since with self-selection smallholders with most to gain would be the ones most likely to enter contracts. That is, smaller more constrained enterprises that were not doing well in the spot market system would have strong incentives to negotiate contracts.

Alternatively, if selection is by agribusiness firms, larger, less constrained smallholders with lower unit costs and less risk exposure could be the most attractive partners. When agribusiness firms have the power to pick and choose smallholders for the contract there are a number of selection factors that seem to be common across contracts. These include previous farming experience of the smallholder, farm size, fertility of farms and community considerations.

Producers in Central and South America with previous experience in growing particular crops achieved good contract performance while other growers who lacked such experience, were not so successful. In Indonesia, a ginger contract was successful from an agronomic viewpoint because the type of ginger required was a juvenile form of a type contractors had grown before. Smallholder farm size is also important in selection of contractees. CDC, Runsten, and Little and Watts argue larger growers are likely to be more attractive partners.

Larger growers can undertake more production hence overheads associated with the contract are a smaller proportion of total costs. This means costs incurred by the firm for provision of extension information and farm visits, purchase of equipment and other capital outlays associated with set-up are lower per unit of contracted output. Large growers are better positioned to bear crop risk, may already possess expertise in crop husbandry and labour management and often have storage and transport facilities. Finally, large growers are better able to manage quality, their produce is less likely to be pooled hence is more easily traced if quality problems occur and they can achieve economies to size in audit and record keeping. These size advantages allow larger growers to be paid lower prices than smaller growers with lower contractual risk for the agribusiness firm.

Farmers with more fertile farms can achieve lower unit costs and hence higher margins from contracts and so can be paid lower prices than smallholders on less fertile farms. In addition, more fertile farms are less risky in terms of yield and their owners are more likely to meet delivery requirements. On the negative side, more fertile farms have higher opportunity value and hence there may be greater risk of the smallholder dropping the contract for some other type of production.

There are logistical advantages for an agribusiness firm selecting farmers from a single geographical region. Most contracts involve visits to smallholders, village headmen and farm group meetings and money is saved if these people are in close proximity both to each other and to processing facilities. In Indonesia contracts often involved one or a few villages in regional clusters. Hence, from the standpoint of the agribusiness firm seeking to establish a new contract, the problem is not simply to identify suitable growers but also a suitable community environment. Contracts need to be harmoniously integrated into communities without creating too much competition for resources resulting in resentment of the contracting firm or its contractees.

Contract default

Enforcement of contracts is central to contract success. A major problem is that agribusinesses are hampered by limited legal recourse when things 'go wrong'. Developing country smallholders are unlikely to use land as collateral on advances of credit or other farm inputs and the traditional nature of land titles and slow court processes means liens on small parcels of land are virtually useless for recovery of small debts. Thus contracted smallholders may either divert inputs provided by the agribusiness firm to other on-farm end-uses or sell them or may divert contracted production to other purchasers without facing the types of penalties imposed on developed country contract farmers who default.

Thus, an immediate problem for the agribusiness firm setting up a contract is to ensure contract default is minimised. From the agribusiness firm's perspective, the major element in contract compliance is providing the smallholder with credible prospects for, and a desire for, contract renewal. In this regard, smallholders can be seen as collateralising future income rather than assets to secure upfront transfers from contracts. The contract must be sufficiently attractive so costs to the smallholder of default related to exclusion in future seasons exceeds benefits from default such as being able to pocket forward payments. If a contract is only marginally attractive in terms of profit then default risk is higher.

Firms can do other things to minimise default. Loans can be channelled through farm groups so that default, providing the group does not collectively default, results in social costs. Monitoring and communication may result in relationships that create personal pressures for contract compliance. Other strategies for avoiding default include having strict rules for dealing with defaulters and, if possible, co-operation between buyers in boycotting diverted production.

Conflict resolution

Contract managers need to be able to deal with disagreements if contracts are to be successful. Glover & Kusterer report many misunderstandings and disagreements arising in contracts and our experience in Indonesia was that smallholder issues over operational aspects of contracts seemed the rule rather than the exception. As contracts mature past an initial 'honeymoon' period, issues may arise about real or imagined wrongs suffered by smallholders from actions by the contracting firm.

Case-studies of contracted Canadian and Australian potato contracts indicate this is also common in developed country contract farming. It seems likely that a 'honeymoon' period after the establishment of a contract is followed by a period of contract tightening. Although there are no published data on prices and effective rates of subsidisation over the lifecycle of a contract, two reasons can be conjectured why contract terms and conditions might tighten as contracts mature. First, first season contracts are likely to include some subsidisation of smallholder setup costs.

This may take the form of intensive attention from the firm's extension officers or assistance with capital investments that is implicitly included in contracts through generously set prices and subsidisation of inputs. The second reason contracts may tighten up is that relatively high cost producers who make lower profits from the contract are less likely to renew contracts or to be asked to renew by the agribusiness firm. As high cost producers leave the contract the agribusiness firm can negotiate tougher terms without jeopardising its procurement base. If this occurs it would leave smallholders with the correct perception that the contract was getting tougher over time and, presumably, that they had something to complain about.

7

Contract Farming and Poverty Reduction

Poverty in Asia is largely a rural phenomenon: although the magnitudes and ratios would vary across countries, almost everywhere in the region income poverty is disproportionately higher in rural areas.

Rural areas are generally worse off in almost every other aspect of poverty and deprivation: people in rural areas tend to have lower levels of health and education; they are more likely to have limited access to basic services such as water and sanitation; and paradoxically, despite depending on agriculture as their main source of livelihood, they also suffer the most from hunger and food insecurity.

The rural poor are also the most affected by environmental degradation: fifty percent of the poor in Asia are found in fragile ecosystems and vulnerable areas, and their livelihoods largely depend on the available natural resources. Environmental degradation reduces the income of the rural poor disproportionately in comparison with the general population. Perversely, as the incomes of the poor shrink, they are left with little recourse but to deplete environmental resources even further, thus precipitating a vicious cycle of ever-worsening poverty and environmental deterioration.

Contract farming is emerging as a promising tool to facilitate transformation from subsistence to commercial agriculture. It also facilitates adoption of high-value crops such as organically grown crops.

Case Study of Thailand

Methodology

This section employs the profit frontier methodology to assess the profitability and profit efficiency of the sampled Thai rice farmers. Profit efficiency is defined here as the ratio of the observed profit to the potential maximum attainable profit. While profit provides a direct measure of relative competitiveness of one type of farm (that is a contract farm) in relation to others (that is a non-contract farm), the concept of profit efficiency can also be useful as an indicator of relative competitiveness. We also attempt to account for selection bias using a two-stage switching regression model. The estimated models are used for subsequent 'counterfactual' simulations of profit and profit efficiency.

The analysis aims to test the following hypotheses:

1. Contract rice farmers are more profitable than non-contract rice farmers for comparable scales of operation; and
2. Contract rice farmers are more (profit) efficient than non-contract rice farmers for comparable scales of operation.
3. Contract farming is biased against small farmers.

Since all contract rice farmers in the sample are certified organic or in transition to becoming organic farmers and all the non-contract farmers are conventional rice farmers, the analysis also throws some light on the debate concerning organic versus conventional agriculture. However the evidence on this must be interpreted with care and it is difficult to draw firm conclusions. This is partly because we cannot separate the effects of an institutional arrangement (a contract) from a technology (organic farming practices) as the contract farming group is influenced by both. Further not all contract farmers are certified organic farmers who have completed the required three year transition period, although we can distinguish between the pure (or certified) organic farmers from those who are either in transition or just starting out to adopt organic practices.

Efficiency and profit frontiers

Efficiency and inefficiency can generally be measured by its components - technical, cost, revenue and profit. Technical efficiency refers to a farm's ability to produce the maximum outputs for a given set of inputs and technology. Or conversely, it can be measured as the farm's ability to utilise the minimum amount of inputs to produce a desirable set of outputs for a given technology. Cost efficiency refers to the ability of the farm to minimise the expenditures required to produce a desirable set of outputs, given their respective input prices and production technology. Misallocation of inputs contributes to cost inefficiency and is sometimes refers to as input allocative inefficiency.

Revenue efficiency refers to the farmer's ability in allocating their outputs in a revenue-maximising manner for a given set of output prices. Finally, profit efficiency refers to a farm's ability to obtain maximum profit for a given set of input prices, output prices, and technology. While technical, cost, and revenue efficiency are necessary for the achievement of profit efficiency, they are collectively not sufficient for profit efficiency. Profit efficiency further requires that technical, cost and revenue efficiency be achieved at the proper scale, that is it requires some kind of scale efficiency.

Here we utilise a dual variable profit frontier, which portrays the maximum variable profit (defined as gross revenue less variable cost) obtainable by a farm given the prices of inputs and outputs, the production technology, and the presence of fixed inputs such as land and capital. The variable profit frontier is more appropriate when farms do not have the flexibility to adjust all inputs. Farms operating on the profit frontier are profit efficient while farms operating under the profit frontier are profit inefficient.

Other studies using profit frontier analysis

In terms of the wider literature, while rice is perhaps the most studied agricultural commodity by researchers, very few have used profit frontiers, which could be due to lack of appropriate data. In a review article by Bravo-Ureta and Pinheiro on efficiency analysis of developing country agriculture, 13 out of 20 studies were on rice farming. However, only two studies used the dual profit frontier approach and only one was on rice farming. Ali and Flinn used a single equation dual profit frontier to examine the efficiency of 120 rice producers from the Punjab in Pakistan. They found that the average inefficiency was 31%. Education was found to have a significant role in reducing profit inefficiency, while off-farm employment and difficulties in securing credit to purchase fertiliser tended to increase profit inefficiency. The other study by Bailey et al. is on dairy farms.

Since 1993 a few more studies have employed profit frontiers. Abdulai and Huffman used a stochastic translog profit frontier to examine the efficiency of 256 farmers in the Northern region of Ghana. They found that the average inefficiency was 27.4%. Their inefficiency analysis suggested that the education of the household head, access to credit, greater specialisation, and location in districts with better access to extension services and better infrastructure were significant variables for increasing profit efficiency. On the other hand, increasing participation in nonfarm activities by farmers and being older tended to lower profit efficiency. Rahman also used a dual profit translog frontier to investigate the efficiency of 380 farms, which produced a modern variety of rice in three agro-ecological regions of Bangladesh. He found that the average inefficiency was about 23%. Farmers with more experience in growing modern varieties of rice, better access to input markets and extension services, located in fertile regions, as well as those with less off-farm work and who owned their land were found to be more efficient.

To our knowledge, there are no other efficiency studies on rice farming which employ the stochastic profit frontier approach. However, there are several efficiency studies of other agricultural products using the stochastic frontier approach since the 1993 review article by Bravo-Ureta and Pinheiro. Arajuo and Bonjean used a stochastic profit frontier to study the efficiency of different land tenure patterns in Brazilian farms. Bhattacharyya and Glover also employed a stochastic profit frontier to examine the efficiency of small versus large farms in India. Wang et al developed a shadow-price profit frontier model to examine the efficiency of Chinese rural households in farming operations. Delgado et al employed the profit frontier approach to investigate the efficiency of large versus small and contract versus independent livestock farms in the Philippines, India, Thailand, and Brazil.

Thai Data

In 2003, the Ministry of Agriculture and Agriculture Cooperatives of Thailand commissioned a survey to investigate the potential of developing organic rice in Thailand for export. The survey covers 5 provinces, 2 of which are in the Northern region namely Phayao and Chiang rai, while the other 3 are in the Northeastern region, covering Ubon Ratchathani, Surin, and Yasothon.

The sampled farms in Northeast Thailand are ones, which practiced conventional agriculture using high level of agro-chemicals until the early 1980s. In the mid-1980s, in Surin and Yasothon Provinces, the movement of natural agricultural practices and environmental conservation promoted by religious groups and several Non-Government Organisations (NGOs) initiated contract farming of organic rice as a solution to the problems faced by farmers. This development was then supported by European NGOs, who wished to produce organic rice for export to their own Fair Trade Networks.

In Ubon Ratchathani, contract farming of organic rice was supported by a semi-NGO as part of its strategy for community development and poverty reduction. It was also an income earning opportunity for the NGO. In contrast organic rice farming in the Northern region was a private sector-led initiative, prompted by demand growth in European countries. In search for land where chemicals has not been applied for organic production, the firms searched for marginal forest areas and introduced rice contract farming to farmers.

The farm survey was conducted during 2002 and 2003 with an updated interview with key informants during early 2005. Hence the monetary data are in prices of 2002-3. Within each province approximately the same number of contract and non-contract farmers were surveyed in the same locality. All of the contract rice farmers are organic or low-chemical farmers while all the non-contract farmers are conventional rice farmers. This resulted in 83 contract-organic and 85 conventional farmers surveyed in the Northern region, and 140 contract-organic and 137 conventional farmers in the Northeastern region. Thus, there are a total of 445 farms surveyed, 168 in the Northern region and 277 in the Northeast.

The contract-organic farms in the survey are categorised into three groups according to the length of their organic farming experience and the restrictions on their farming practices. Farmers in the 'certified' organic group include those certified to avoid totally the use of chemical fertiliser, pesticides or herbicides. They are mostly more than four years into organic farming. The 'transitional' organic farming group represents farms, which were under transition (mostly two-four years into organic farming), while the 'initial' organic farming group represents farms, which had just gone organic.

Those in the 'transitional' organic or 'initial organic groups in principle should be committed to stop using chemical fertilisers, pesticides and herbicides. The fact that some still do is likely to be due to an ineffective inspection and certification system. In this sample all organic farmers are contract farmers, while all conventional farmers are non-contract farmers. All certified and transitional organic farmers received a premium price based on their years of organic practice.

However not all contract farmers are pure organic farmers making it impossible to completely merge the organic and contract groups.

Information on quantity and value of rice output and major inputs were gathered in the farm survey. The major inputs include seed, hired and family labour, chemical fertiliser, organic fertiliser, pesticides and herbicides, fuel, machinery rental, land, and capital assets. In addition, data were collected on the characteristics of farmers and farms. Table 1 provides a summary of the characteristics of the sampled farms by contract and non-contract farmers and by region. Overall, household heads for contract farms were significantly younger (age of 49 versus 51 years) and better educated (2.86 versus 2.36 years of formal education) than for non-contract farms. These differences were also true for the two regions except that there was no significant difference between the age of the household heads between contract and non-contract farms in the Northern region. With respect to experience in organic rice farming, contract farmers in the North had a significantly higher level than in the Northeast (5.83 versus 3.23 years).

Table 1. Characteristics of sampled farms

	Total Sample Means	Non-Contract Farmers Means	Contract Farmers Means	pvalue
North (No. of farms)	168	85	83	
Age of household head	47.90	47.56	48.24	0.6660
Education of household head (years)	2.52	2.32	2.72	0.0229
Number of household members engaged in rice farming	2.14	2.15	2.13	0.8625
Female members engaged				

in rice farming (%)	0.48	0.50	0.46	0.5890
Land allocated to rice	11.98	10.23	13.77	0.0044
Land ownership (%)	85%	76%	94%	0.0013
Years in rice farming	35.47	32.21	38.81	0.0858
Years in fragrant rice farming	10.45	8.65	12.29	0.0164
Years in organic rice farming	2.88	0.00	5.83	0.0000
% of saline soil	5%	6%	4%	0.4931
% of income from non-agricultural activities	8%	7%	9%	0.7390
% of agricultural income from rice farming	58%	56%	60%	0.5650
% of labour from family	66%	67%	65%	0.6900
% of seed from own supply	42%	38%	47%	0.2229
% of organic fertiliser from own supply	15%	4%	26%	0.0000
NORTHEAST	277	137	140	
Age of household head	50.72	52.81	48.68	0.0019
Education of household head (years)	2.67	2.39	2.94	0.0004
Number of household members engaged in rice farming	2.53	2.54	2.51	0.8482
Female members engaged in rice farming (%)	0.54	0.54	0.53	0.9690
Land allocated to rice	12.17	11.01	13.31	0.0169
Land ownership (%)	93%	92%	94%	0.4478
Years in rice farming	43.00	42.45	43.54	0.6747
Years in fragrant rice farming	13.97	16.50	11.49	0.0001
Years in organic rice farming	1.63	0.00	3.23	0.0000
% of saline soil	8%	9%	7%	0.6204
% of income from non-agricultural activities	14%	16%	12%	0.2790
% of agricultural income				

from rice farming	62%	61%	64%	0.5320
% of labour from family	76%	76%	76%	0.8783
% of seed from own supply	65%	70%	59%	0.0588
% of organic fertiliser from own supply	56%	53%	60%	0.1223
TOTAL	445	222	223	
Age of household head	49.66	50.80	48.52	0.0261
Education of household head (years)	2.61	2.36	2.86	0.0000
Number of household members engaged in rice farming	2.38	2.39	2.37	0.8384
Female members engaged in rice farming (%)	0.51	0.52	0.51	0.7310
Land allocated to rice	12.10	10.71	13.48	0.0003
Land ownership (%)	90%	86%	94%	0.0040
Years in rice farming	40.16	38.53	41.78	0.1378
Years in fragrant rice farming	12.64	13.50	11.79	0.0093
Years in organic rice farming	2.10	0.00	4.20	0.0000
% of saline soil	7%	8%	6%	0.4430
% of income from non-agricultural activities	12%	13%	11%	0.4650
% of agricultural income from rice farming	61%	59%	63%	0.3920
% of labour from family	72%	73%	72%	0.7475
% of seed from own supply	56%	58%	55%	0.5299
% of organic fertiliser from own supply	41%	34%	48%	0.0005

The average number of household members engaged in rice farming was very similar for contract and non-contract farmers (2.37 versus 2.39 persons) overall and in both regions. The percentage of female members engaged in rice farming was also very similar for contract and non-contract farmers (51% versus 52%) overall and in both regions. Overall, contract farmers allocated an average of 13.48 rai of land to rice farming, which was significantly higher than the non-contract farmers who allocated only 10.71 rai. Similar patterns were also exhibited in both regions.

With respect to land ownership, while there was no significant difference between contract and non-contract farmers in the Northeast, contract farmers in the North had a higher percentage of ownership than the non-contract farmers (94% vs. 76%). While the percentage of income derived from non-agricultural activities was significantly lower for the farms in the North (8%) than in the Northeast (14%), there were no significant differences between the two contracting groups within each region. As to the percentage of agricultural income derived from rice farming, there were no significant differences between regions and within the two contracting groups.

While farms in the Northeastern region used a significantly higher percentage of family labour (76%) when compared to the North (66%), there were no significant differences between contract and non-contract farmers within each region. While the contract and noncontract farmers in the North on the average showed no significant differences in using their own seeds, non-contract farmers in the Northeastern used significantly more seeds from their own supply than contract farmers (70% versus. 59%). In terms of utilisation of organic fertilisers, contract and non-contract farmers in the North were similar in the percentage from their own production (slightly more than half).

However, contract farmers in the Northeast used significantly more organic fertiliser derived from own supply than non-contract farmers (26% versus 4%). Finally, both regions had a similar endowment of saline soil and there were no significant differences in soil distribution between contract and non-contract farmers. Since contract and non-contract farming tend to have different production processes, we estimate their profit efficiency separately.

Results: Comparative Profitability

Since our interest is in the estimation of profit inefficiency and its determinants, we do not focus on the estimates of the stochastic profit frontier except for the derived profit elasticities. Table 2 shows the profit elasticities with respect to the prices of the six variable inputs and the two fixed factors for both contract and non-contract farms. The profit elasticities of contract farms with respect to seed price, wage and energy are negative as expected yet not statistically significant, while the elasticities with respect to the prices of chemical fertiliser, organic fertiliser and machinery are positive yet insignificant.

As to the non-contract farms, the profit elasticities with respect to all the input prices are of the correct sign except for seed, which is also not statistically significant. For both contract and non-contract farms, profit elasticities with respect to the two fixed factors (land and capital) are also of the right sign but capital is not statistically significant. The estimated profit elasticities with respect to land are 0.87 for contract farms and 0.98 for non-contract farms, indicating that profit tends to increase by less than 1% with a 1% increase in land allocated to contract or non-contract rice farming.

Table 2. Estimated profit elasticities for farms

Profit elasticity with respect to	*Contract farms*		*Non-contract farms*	
	Elasticity	*p-value*	*Elasticity*	*p-value*
Variable inputs				

Seed price	-0.242	0.151	0.100	0.414
Wage	-0.076	0.277	-0.017	0.884
Chemical fertiliser price	0.106	0.653	-0.012	0.962
Organic fertiliser price	0.021	0.547	-0.171	0.002
Machinery	0.018	0.447	-0.047	0.108
Energy	-0.052	0.119	-0.001	0.985
Fixed inputs				
Land	0.868	0.000	0.975	0.000
Capital	0.006	0.784	0.027	0.385

We first test the hypothesis that 'contract rice farmers are more profitable than noncontract rice farmers for comparable scales of operation.' Here we focus on two distinct measures of profit one deducting only direct cash costs from sales revenue ('profit over cash costs') and the other deducting both cash and imputed non-cash costs ('profit over total variable costs'). We place most emphasis on profit per unit of land (baht per rai).

Contract farmers on the average generated a profit over total variable cost of 1,234 baht per rai in the North and 1,098 baht per rai in the Northeast. On the other hand, non-contract farmers produced a profit over total variable cost of 731 baht per rai in the North and only 273 baht per rai in the Northeast. Differences in profitability are less sharp when costs include only cash costs excluding the imputed value of own inputs, like family labour and seeds ('profits over cash costs'). Differences in profitability can largely be explained by the significantly higher price of rice received by the contract farmers (6.5. versus 6.0 baht/kg in the North and 7.9 versus 5.9 baht/kg in the Northeast).

The marked difference in price for organic rice between the two regions is explained by the different price formulae used in private sector-based contract farming in the North and an NGO-based system in the Northeast. In the North the contracting firms offered a fixed margin of 0.5 baht above the market price of conventional rice at harvesting. In the North east the price was fixed at the start of the season based on negotiations between the NGO and the farmers. On the other hand, it is interesting to note that yield in kg per rai or was very similar for the contract and non-contract farmers in both regions. Average yields are considerably lower in the Northeast, however, due to a higher level land degradation.

In terms of the role of organic farming practices it is important to note that while contract farmers in the Northeast, contracted to an NGO with broader social objectives, appeared wholly organic with zero expenditure on chemical fertiliser, pesticides and herbicides, the transition and initial organic groups in the North continued to use them, due to an ineffective monitoring system operated by the contracting firms, although at lower levels per rai than non-contract farms.

For example expenditure on chemical fertiliser per rai in the initial organic group was roughly two-thirds of that for conventional farms. Furthermore, organic farmers in the Northeast used more on-farm organic fertiliser than the contract farmers in the North. Invested capital assets (valued at baht per rai) were significantly higher for contract farmers in the North, while there was no significant difference between contract and non-contract farmers in the Northeast. Farmers in the North were generally far more capitalised than the Northeast.

While organic farmers in the North regardless of their stage of transition achieved similar levels of profit (in terms of profit over cash cost per rai), the certified organic farmers in the Northeast were considerably more profitable than the transitory and initial organic farmers. This is despite the fact that in the North, as we have just noted, there were considerable differences in terms of organic practices between the three different groups, with only the certified group being wholly organic.

In the Northeast where profitability was generally lower than the North, the profitability of the initial organic farmers (defined as profit over cash expenditure), who continued to use chemical fertilisers, was roughly 25% above that of conventional farms. This profitability pattern can again be largely explained by the price of rice received by the farmers. While the price of rice was not significantly different among the three organic groups in the North, the price received by the certified organic farmers (10 baht per kg) in the Northeast was considerably higher than that received by the transitory and initial organic farmers (7.1 and 6.3 baht per kg respectively) and nearly double that received by non-contract farmers.

Table 3. Profitability by farm size (profit after cash costs per rai)

Land category	All farms	Non-contract farmers	Contract farmers	p-value
0-5 rai	1,719	1,374	2,432	0.0000
6-10 rai	1,744	1,413	2,076	0.0000
11-20 rai	1,723	1,337	2,021	0.0000
>20 rai	1,646	1,276	1,866	0.0057
Total	1,721	1,369	2,072	0.0000

Table 3 relates profit to farm size. Profit after cash costs for contract farmers per unit of land decreases with the increase in farm size, while for non-contract farmers profit after cash cost is more stable. We find no support for our third hypothesis and conclude from this that contract farming as practiced in these areas of Thailand does not seem to be biased against smaller farms in terms of profitability, as is sometimes argued. Furthermore, for all farm sizes profits are significantly higher for contract farmers, as compared with non-contract farmers.

Selection bias and counterfactual simulation

The above profitability comparison reveals that contract farms in the sample generally have higher profits than non-contract farms. However, this profitability difference does not necessarily indicate that contracting has a positive impact on profits because it could be caused by selection bias. That is, the higher profitability in contract farming may merely reflect the fact that farms with the potential of securing higher profitability are more likely to become contract farms. In other words, these contract farmers might have relatively high profits whether engaging in contract or non-contract farming. A counterfactual simulation can help sort out the impact of contracting on profitability. In brief, the key to this approach is to estimate farms' counterfactual profits and compare these to their actual profits.

The counterfactual profit of a contract farm is defined as the hypothetical profit that it could have earned had it farmed like a (typical) non-contract farm. Similarly, the counterfactual profitability of a non-contract farm is defined as the hypothetical profit that it could have earned had it farmed like a (typical) contract farm. Non-contract farms in the sample generally sold their rice at lower prices than contract farms. We use the rice prices of contract (or non-contract) farms in the estimation of the counterfactual profits of non-contract (or contract) farms. Higher actual than counterfactual profits for contract farms would indicate that contract farms would have been less profitable had they operated like a non-contract farm. Similarly, lower actual than counterfactual profits for non-contract farms would indicate that non-contract farms would have been more profitable had they operated like a contract farms. The counterfactual results are given in table 4.

Had contract farms operated like a non-contract farm, their counterfactual profits would (on average) have been 31% lower than their actual profits; the differences are 49% and 21% respectively for contract farms in the North and Northeast. Conversely, had non-contract farms operated like a contract farm, their counterfactual profits would have been 47% higher than their actual profits; the differences are 9.4% (significant at 10% level) in the North and 72% in the Northeast. These results clarify that the observed higher profitability in contract farming is not simply because of contract farming attracting the more profitable farms; rather, it is evidence supporting the hypothesis that contract farming tends to be more profitable than non-contract farming.

Table 4. Counterfactual versus actual profits

Profit	*All* Profit differences	*All* p value	*North* Profit differences	*North* p value	*Northeast* Profit differences	*Northeast* p value
Contract counter factual vs. Contract actual	31.4%	0.0000	-48.8%	0.0000	-21.1%	0.0059
Non-contract counterfactual vs. Non-contract actual	47.4%	0.0000	9.4%	0.0957	71.8%	0.0000

Results: Comparative Profit Efficiency

Here we test the second hypothesis that 'contract rice farmers are more profit efficient than non-contract rice farmers for comparable scales of operation.' Table 5 shows the profit efficiency, actual profit and profit loss per rai for contract and non-contract farms by region. Profits here are after deducting cash costs only from sales revenue. Profit loss is defined as the amount of unrealised profit due to inefficiency and can be calculated as the difference between maximum possible profit (that is, profit on the profit frontier) for each farm and its actual profit. The estimated mean profit efficiency score for the entire sample farms is 0.68. In other words, significant profit inefficiency occurred among the sample rice farms in Thailand and farms could increase their profit by 32% or 842 baht per rai by improving their efficiency.

Table 5. Profit efficiency of contract versus non-contract rice farmers

	N	*Actual Profit (baht/rai)*	*Profit Loss (baht/rai)*	*Profit Efficiency index*
All	443	1,721	842	0.68
Contract	222	2,072	906	0.72
Non-contract	221	1,369	778	0.64
p-value		0.0000	0.0388	0.0032
North	168	1,847	650	0.76
Contract	83	2,001	727	0.76
Non-contract	85	1,697	575	0.76
p-value	0.0047	0.0934	0.9916	
Northeast	275	1,644	960	0.63
Contract	139	2,114	1,014	0.69
Non-contract	136	1,163	905	0.56
p-value		0.0000	0.1769	0.0002

As shown in Table 5 farmers in the North, where new land was brought into production, exhibited significantly higher profit efficiency than farmers working on the more degraded land of the Northeast, with a mean efficiency of 0.76 versus 0.63. Overall, contract farmers were significantly more profit efficient than non-contract farmers, with a mean profit efficiency of 0.72 versus 0.64. This is also true for farmers in the Northeast where contract farmers are found to be significantly more profit efficient than non-contract farmers (0.69 versus 0.56). However, the efficiency scores of contract and non-contract farmers in the North were virtually the same on average, although the scores were more diverse among the non-contract farmers.

Table 6 shows the profit efficiency across different farm sizes for contract and noncontract farmers. Similar to profitability, contract farmers had higher profit efficiency for all farm sizes except those greater than 20 rai. Contract farmers appear to show a slight tendency to decreasing profit efficiency for larger farm sizes, while non-contract farmers are more homogeneous across all farm sizes. Similar to the profitability, comparison by farm size, with respect to profit efficiency contract farming does not seem to be biased against smaller farms.

Table 6. Profit efficiency by farm size

Farm Size	*0-5 rai*	*6-10 rai*	*11-20 rai*	*>20 rai*
All Farms				
Actual profit (per rai)	1,719	1,744	1,723	1,646
Profit loss (per rai)	821	774	850	1,067
Profit Efficiency	0.69	0.70	0.67	0.64
Non-Contract Farms				
Actual profit (per rai)	1,374	1,413	1,337	1,276

Profit loss (per rai)	801	762	764	818
Profit Efficiency	0.64	0.65	0.62	0.64
Contract Farms				
Actual profit (per rai)	2,432	2,076	2,021	1,866
Profit loss (per rai)	862	786	916	1,215
Profit Efficiency	0.78	0.75	0.70a	0.64
p-value of profit efficiency between contract and non-contract farmers	0.0276	0.0325	0.1351	0.9902

Table 7 shows profit efficiency among the different groups of organic farms we have identified. Farmers with a longer history and more experience in organic farming (the 'certified' group) appear to be more profit efficient, as well as more profitable. However, multiple range tests show that all three groups of organic farmers in the North exhibited similar profit efficiency as well as profitability. In fact, in terms of profit efficiency they were not different from the conventional non-contract farmers.

Table 7. Profit efficiency by different stages of organic farming

	Certified Organic	*Transitory Organic*	*Initial Organic*	*All Organic*	*Con-ventional*
All Farms					
Actual profit	2,428	1,901	1,752	2,072	1,369
Profit loss (per rai)	956	904	790	906	778
Profit Efficiency	0.75	0.71	0.70	0.72	0.64
North					
Actual profit (per rai)	2,018	2,042	1,927	2,001	1,697
Profit loss (per rai)	745	691	727	727	575
Profit Efficiency	0.77	0.77	0.75	0.76	0.76
Northeast					

Actual profit (per rai)	2,849	1,867	1,416	2,114	1,163
Profit loss (per rai)	1,172	955	909	1,014	905
Profit Efficiency	0.73	0.69	0.60	0.69	0.56

In the Northeast profit efficiency was not statistically different between the certified and transitory groups, although it was higher for these than for the initial organic group, whose efficiency was statistically similar to that of conventional non-contract farmers.

Counterfactual simulation for profit efficiency

Similar to the case of the actual-counterfactual profitability comparison, the difference in profit efficiency between contract and non-contract farming can also be evaluated through comparing actual and counterfactual efficiency. The methodology is similar to that used in estimating the counterfactual profitability. To estimate the counterfactual efficiency of a contract farm (that is its profit efficiency when hypothetically operating like a non-contract farm), the first step is to use the estimated profit frontier of non-contract farming to estimate the maximum profit the contract farm would have obtained had it produced like a noncontract farm with 100 percent efficiency. The second step is to use its hypothetical profit estimated from the counterfactual profit simulation to represent its counterfactual profit in non-contract farming. Then the difference between this counterfactual profit and the counterfactual frontier can be used to measure the farm's counterfactual efficiency. The counterfactual efficiency of a non-contract farm can be estimated similarly.

Table 8 shows that contract farms in the entire sample would not have had very different counterfactual efficiency from their actual efficiency (69% versus 70%)

had they operated like a non-contract farm. This mainly reflects the situation in the Northeast, while contract farms in the North would have reduced their efficiency from 74% to 68% by counterfactually operating like a non-contract farm. With respect to the non-contract farms, generally for the entire sample, non-contract farms would have had a slightly higher counterfactual than actual efficiency (69% versus. 66%), and the difference is statistically significant at 10%. Again, this mainly reflects the situation in the Northeast (68% versus 59%), while surprisingly the non-contract farms in the North would have had lower counterfactual efficiency than their actual efficiency (69% versus. 77%).

Table 8. Counterfactual versus actual efficiency

Counterfactual versus actual	Average profit efficiency		
Contract farming	*Entire*	*North*	*Northeast*
Contract counterfactual	0.6879	0.6798	0.6928
Contract actual	0.6988	0.7409	0.6736
p-value	0.4965	0.0326	0.3096
Non-contract farming	*Entire*	*North*	*Northeast*
Non-contract counterfactual	0.6854	0.6873	0.6841
Non-contract actual	0.6596	0.7658	0.5913
p-value	0.1284	0.0009	0.0000

In short, the results from the counterfactual efficiency estimations are mixed and do not generally support the hypothesis that contract farming enhances profit efficiency. Indeed, the efficiency patterns appear to be different between the North and Northeast regions, perhaps due to different contract management systems and different land endowments.

8

Equity and Efficiency in Contract Farming

Contract farming refers to a system where a central processing or exporting unit purchases the harvests of independent farmers and the terms of the purchase are arranged in advance through contracts. The terms of the contract vary and usually specify how much produce the contractor will buy and what price they will pay for it. The contractor frequently provides credit inputs and technical advice.

Contracting is fundamentally a way of allocating risk between producer and contractor; the former takes the risk of production and the latter the risk of marketing. The allocation of risk is specified in the contract which can vary widely; some agree to trade a certain volume of production; in others the contract specifies price but not amount.

Typical Contracts

— Market specification contracts: future purchase agreements which determine quantity, timing and price of commodities to be sold.

— Resource-providing contracts: specify the sorts of crops to be cultivated, some production practices and the quality and standardisation of the crop through the provision of technical packages and credits.

— Production management contracts: associated with large outgrower and nucleus-estate schemes, directly shape and regulate the production and labour processes of the grower.

The definition of contract farming is often confused because there are so many different types of contracts and actors (private sector firms, public sector firms and parastatals, international aid agencies). The definitions suggested by Ellman and Glover and Kusterer, provide an insight into the types of schemes which exist under the label 'contract farming'.

Outgrower schemes: Schemes that provide production and marketing services to farmers on their own land. For Glover and Kusterer, these generally connote a government scheme with a public enterprise, purchasing crops from farmers, either on its own or as a joint venture with a private firm. Glover and Kusterer also use the term contract farming to refer to the same arrangement in the private sector.

Nucleus Estate-Outgrower Schemes: A core estate and factory is established and farmers in the surrounding area grow crops on part of their own land, which they sell to the factory for processing.

Multipartite Arrangements: A term often used in the literature to emphasise the participation of several actors.

One dominant approach to contract farming, which lies halfway between natural and social sciences, emphasises the technical characteristics of contract commodities as central in determining production relations in agriculture. This commodity-based approach classifies schemes according to whether they produce 'classical' export crops (tea, oil palm, sugar), horticulture or staples. The argument is that each contracted commodity is associated with specific conditions of production and labour regimes, which affect its potential for generating local development.

Commodity specific limits pose both organisational limitations and opportunities which will determine whether they are better grown in outgrower schemes with centralised processing units; or decentralised outgrower schemes etc. Under this school of thought, the success of different trees in outgrower arrangements is primarily determined by their properties and outputs. Technical constraints are doubtlessly an important factor, but there are some apparent problems with this approach, which will be returned to later.

Contract farming should be examined case by case in order to understand its potential as a tool in rural development strategies. Such a study should focus on an analysis of the economic logic behind contract farming and its political implications, in particular the effects of contract growing on the process of empowerment and political organisation by outgrowers. His studies have also paid considerable attention to the impact of contract farming on regional development and those not included in the contracts.

Contract farming has attracted such polarised viewpoints for several reasons. One is that with its dramatic linking of smallholder and national/ international capital, it is bound to excite pre-existing ideological viewpoints of what the outcome of such a marriage will be. Another is that there are many actors in contract farming and most have different motives. There is firstly the economic logic of contract farming as a way to divide risk between growers and contractors. There is also the public motive of promoting smallholder development, transferring technology, building a smallholder political base and generating foreign exchange. Further, smallholders are motivated by, amongst other factors, food security, cash flow and risk avoidance. These private and public objectives, often operating simultaneously in a fluctuating market, can obscure what part of contract farming is being evaluated and how.

Nevertheless, an examination of institutional roles and objectives is valuable for an understanding of how tree outgrowing might work. As Arnold and Dewees argue, it is important to shift the debate away from the 'forestry' and 'welfare' approaches to the analysis of farm trees in the 1970s and 1980s. Policies to promote tree outgrowing will be motivated both by development and profit objectives; smallholders will most likely be considering both cash and food security; and public policies will have to balance consumer and producer prices etc.

Emergence of Contract Farming

There are several global factors that led to the emergence of contract farming. One was the break up of many plantations in colonial countries after Independence when foreign agri-businesses were subject to nationalist pressures, threats of expropriation and new conditions of profitability with a changing international division of labour. In a world of price instability, political uncertainty over nationalisation and tax regimes, unionisation of some labour forces, it is easier for a foreign company to drop short-term contracts with smallholders than the management of an estate.

Second, independent peasants and newly settled pioneer farmers are drawn into outgrower schemes under state and/or private auspices. These schemes are often run in cooperation with international lending agencies and development banks. The International Finance Corporation and the Commonwealth Development Corporation are amongst the biggest supporters of outgrower schemes and have pioneered palm oil, cocoa and rubber contracting across Asia, Latin America and Africa. Little and Watts trace the recent popularity of contract farming to IMF austerity measures and attempts to revive flagging export markets. The World Bank has promoted contract farming as a way of creating dynamic partnerships between private capital and smallholders, which will lead to technology transfer, innovation and market growth.

Most outgrower schemes have more than one sectoral actor involved, especially in Africa where cases of strictly private sector schemes are rare. The usual model for large contract farming schemes is for the state to have some involvement; a review of 67 contract farming schemes in Africa showed that 70% had either full or joint state ownership. Some of the largest outgrower schemes such as palm oil in the Philippines, rubber in Malaysia and tea in Kenya are public sector schemes.

In practice the public-private sector dichotomy is even more complex as a public sector scheme may be under private management; or a private scheme may be supported by subsidies, extension and research. Judgements on the economic logic of the contract vary widely. Neo-classical economists would argue that contracts are freely entered and serve to insulate the small producer from the market. They allow the smallholder to make use of their endowments in imperfect markets and arrive at combinations of income, effort and risk, which reflect their resources and preferences.

Little and Watts would argue that 'contract production among peasant and smallholder growers aims to exploit non-wage household labour through dense networks of dependence and subordination'. Contract farming appears in such diverse forms, that it is better to focus on the motives each participant has and the relationships of contracting parties, rather than trying to isolate a generic institution.

Different Objectives

In many cases, a single contract farming scheme must fulfil the multiple and often conflicting objectives of its various partners. The development component of schemes may require the schemes to meet certain criteria for equity, but development projects are often not economically viable in the long run. In other cases, contractor monopsony and profit maximisation, often combined with a powerful contract, can have serious repercussions for small contract farmers. These problems are all the more difficult to disentangle and evaluate when a scheme has multiple partners with different and sometimes hidden agendas.

Private contractor

Outgrower schemes that allow the company to delegate production to smallholders have several advantages. The contract assures the company of regular inputs of raw material from the smallholders so that it is able to meet its economies of scale. They would not be able to achieve this through purchases on the open market, whereas contracts can specify planting dates as well as total quantity to be delivered. The contract therefore both reduces uncertainty and gives the company control of the production process. Further, the company does not have to invest in land, hire labour or large scale farming operations.

Some companies (and parastatals) retain a 'nucleus estate' surrounded by outgrowers, especially when the economies of scale of the processing plant (such as for palm oil) depend on a certain volume of throughput. Many companies have withdrawn from production completely, delegated responsibility for processing, and retained control of only the most critical stages of marketing. Avoiding conflict over landownership and labour issues is an important advantage of the contract, whatever the level of integration of the company. Little and Watts argue that not only does contract farming allow potential problems with labour to be avoided, but it allows the company to profit from self exploitation and exploitation of the family.

Apart from these economic reasons there are several political reasons why contract farming is attractive to private companies. It allows the company not to invest too many resources in a country and therefore to avoid the risk of appropriation. Contract farming, presented as a smallholder friendly scheme, can be good for the public image of a company and give the impression that it is progressive. This can be exploited by the company to encourage the State, or even international aid agencies to provide credit for operating capital or for the rehabilitation of plantations.

Governments

Contract farming schemes offer governments the

potential of combining development issues with foreign exchange earnings. They are therefore often very politically attractive. They avoid foreign ownership of large tracts of land, and may also create the impression that other features associated with plantations such as the enclave effect are avoided. Outgrower schemes may also appeal to governments who whilst realising the political necessity of addressing the needs of the smallholder, prefer to keep them under a central authority. Ellman has found the stated aims of the government to increase cash and food crop production; maximise rural employment generation; improve social facilities and rural infrastructure often clash with implicit objectives which are to (1) accord political acceptability to a plantation-type project; (2) to mobilise cheap family labour for production; and (3) to move possibly reluctant farmers from a heavily populated area to an under populated one. Outgrower schemes are often linked to, and facilitate resettlement schemes. Further, contracting often creates lucrative opportunities for absentee landlords, especially with crops such as rubber which are not labour intensive.

Glover and Kusterer argue that contracting has trade-offs for policy makers, even if their objectives are solely motivated by promoting rural development. 'The contracting relationship is not a 'zero sum game'; the distribution of benefits between the firm and its growers can affect the total magnitude of benefits available'. For example, a bank may supply credit to growers and expect high producer prices so that the growers can repay their debt; the price control agency will try to keep prices down to safeguard the interests of the consumer, and the Ministry of Finance will expect enough revenue for the company to be able to tax its profits.

Outgrowers

The primary motive for smallholders to become outgrowers is market access, and therefore increased income from the sale of a cash crop with an acceptable level of risk. Local markets for commodities are often highly volatile and prices can drop within days. International markets are more stable than local ones but inaccessible without specific channels such as those provided by outgrower schemes. Smallholders are often reluctant to adopt new technologies because of the risks involved. Further, there is no guaranteed supply of inputs such as fertiliser and agro-chemicals from the government and public extension services that are often poor.

Contract farming provides the smallholder with access to these technologies and with extension, either private or priority treatment from the public extension service. The company has a large vested interest in ensuring the smallholder has access to these services. It also gives them access to credit, which is one of the most frequently given reasons for smallholders to become outgrowers. An analysis of the different systems for providing credit will be returned to in a later section.

Foreign Aid Agencies

Contract farming is attractive to foreign aid agencies for many of the same reasons that it appeals to national governments. It provides a regulated system for channelling large amounts of money straight to smallholders; it provides for the transfer of technology; and it involves different sectors. Many donors, particularly in Africa, prefer private sector involvement. The other motive is profit;

Constitution of the Contract

There are three observations that it is useful to bear in mind in considering the constitution of contracts. The first is that the contract is the means by which risk is distributed between the outgrower and the contractor. The second is that the contract is a 'representation of a relationship rather than the relationship itself' and related to this, the constitution and administration of a contract is highly dependent on the political and economic environment in which it is embedded.

Contracts can be thought of as varying in intensity between those that pay the market price on delivery and those in which every detail, including the production process and price, is fixed. The division of value added between the grower and the contractor is determined by the policy objectives of the scheme; crop characteristics and the dependencies these create; the economic strength of the contractor and outgrower; and the alternative markets available to them. The following table outlines the typical rights and obligations in smallholder contracts.

Table 1. Rights and obligations of farmer and project authority in smallholder contracts

Farmer Obligations	*Project Authority Obligations*
1. Use land for purposes specified in contract	1. Supply credits and inputs
2. Follow production regulations specified in contract	2. Provide technical and managerial support
3. Maintain internal farm roads and drains	3. Maintain infrastructure
4. Sell crop through project authority	4. Purchase all production of acceptable quality

5. Repay loan	5. Pay farmer according to agreed formula
	6. Maintain accounts in a comprehensible form
Farmers rights	*Project authority rights*
1. Timely receipt of services and payments specified as obligations of project authority	1. Timely recovery of payment for services provided to farmers
2. Compensation in the event of default by project authorities on any of obligations	2. Purchase of crop as specified in contract and imposition of penalties in the event of its default

Contracts should in theory specify in detail the rights and obligations between the outgrowers and the contracts, including the penalties for breach of contract by either side. However as Glover has commented, few contracts are 'perfectly contingent' and most are vulnerable to interpretation. Commercial contracts, at least in the initial period, will try to ensure that they have the best possible arrangement for exploiting the resources of their outgrowers. In the initial phases of the project, outgrowers will be attracted by price, credit and technical incentives offered by a company. As will be explored later in more detail, many companies do offer these in the initial phases of the project and then lock outgrowers into production through exploiting gaps in the contract.

Contractors, whether commercial or project, frequently have long waiting lists of outgrowers who would like to join their schemes and are therefore able to be demanding in the requirements. For example, they can often specify the assets that the applicant should have, level of experience with the crop, availability of labour to produce the crop, secure title to land and education. Many of the requirements eliminate poorer farmers, and the selection procedure is also often dependent on interviews and patronage ties. The contracts can also specify age and marital status of an outgrower. For example in Ghana, oil palm outgrowers have to be married with children and be able to direct their family labour to the production of oil palm. The contract, and any debts incurred under it, is inheritable and transferable.

The contract can also specify the amount of land which can or should be put under the contracted crop. For example the Kenyan Tea Development Authority specifies a limit to the area put under tea in order to ensure that the labour to harvest it is sufficient and the output maintains a high quality. More commonly contracts enforce monocropping. They either ban the cultivation of food crops entirely or specify how much can be planted and where. This is a major potential problem with contract farming. In some areas, for example some Tanzanian highlands where tea has replaced maize, the contracted crop is replacing an uneconomic subsistence crop and the overall welfare effects are positive. However, in many areas this regulation leads to spiralling food prices. More commonly, smallholders continue to plant subsistence crops, although without any extension support or guidelines, which contributes to suboptimal production of both crops.

Most problems which are incurred under contract are a consequence of the contingent nature of the contract, and manipulation on both sides, rather than the terms which are written into the contract. The relative dependencies of the contractor and outgrower are extremely important in deciding how contracts are administered. The most common problem for the contractor is that they are unable to maintain their monopoly of the market and other buyers appear and offer a better price. The contract usually safeguards the contractor from such eventualities by specifying levels of production, deducting costs in advance and supplying credit. A further problem is that another contractor moves into the area and takes advantage of the investment the pioneer company has injected into technical know-how and infrastructure development to offer outgrowers a better deal.

Companies can also face problems from manipulation of the product, for example, adding stones for weight, adulterating produce as revenge, and using patronage ties to upgrade produce and to divert inputs intended for contracted crops elsewhere. One of the most common manipulations of the contract relates to quality standards which the firm can arbitrarily raise to control volume and to get a proportion of the produce at a lower price. Many companies specify production procedures which maximise short term output and mine the soil at the cost of long term sustainability.

Companies are also able to levy unspecified charges for goods and services delivered and deduct these directly from payments for goods. Some companies will argue, for example, that an outgrower has not weeded their field properly and then send a labour team around to do this and charge the cost to the outgrower. In fact, one of the most common complaints in all the reports relate to unspecified charges and late payments.

Glover has pointed out that it is not always clear whether a breach of contract is deliberate or caused by bad coordination. For example, it is difficult to coordinate collection and delivery to many smallholders over a different area and when a truck is late, it is possible that this is not deliberate, although it may mean the produce has spoiled and gets a lower price. There are other manipulative practices which are not company procedure but the operation of corrupt officials. It is easy for officials to exploit tea outgrowers who need inputs or early assistance with extension, to bribe oil palm outgrowers whose produce is dependent on being processed, and to charge a routine fee to ensure that coffee is placed in the right grade.

Land Tenure

The outgrowers in settler schemes have to sign more stringent contracts than smallholders who own their land. Smallholders are generally free to cultivate what they want using whatever method of cultivation they desire. This was not the case prior to Independence when 'cultivation by regulation' was central to colonial agricultural policy and smallholders were barred from cultivation plantation crops.

The only way in which cultivation can be regulated when land is privately owned, is through the contract which obliges farmers to follow the rules of cultivation in return for credit, inputs and marketing. Because smallholders have to be induced to become outgrowers, the terms, at least initially, have to offer the smallholders some economic incentive. The land in settler schemes is often leased in short tenures and the lease contingent on following strict production procedures. The contractor therefore has much more control over cultivation, the terms of the contract, and the extent to which they are able to manipulate the contract.

Some contracts have been given for contract farming on communally owned land. For example, in Nigeria the Risonpalm Nucleus Estate does not enter into production contracts with outgrowers, but instead contracts with middlemen for the supply of palm fruit from communally owned palm groves and so avoids the need to disturb traditional land tenure.

Comparative Advantage of Different Sectors

Government departments: The government is difficult to bypass because it possesses obvious economies of scale. It is the biggest employer of trained agricultural personnel and drawing government staff away through awarding high salaries has negative long term consequences. However, a degree of financial and managerial autonomy is often critical in allowing a government department to overcome some of the bureaucratic difficulties, which could slow a project down.

Parastatal authorities: Parastatals have the potential of being a compromise between government departments and private sector companies, but in practice they often run into similar shortcomings to those which characterise government departments. They are not commercially viable, they are subject to political interference and manipulation, and they are bureaucratic. However, some parastatals which have some autonomy perform well.

Private or joint venture companies: Their advantage lies in managerial and financial autonomy, accountability and high quality staff. However there are possible conflicts of interest in a commercial company between shareholders and producers. These can however be minimised if the growers own a stake in the company, if they are protected by the government, or if there is a separate organisation which protects their interests. Private sector companies also tend to adopt authoritarian styles of management which do not encourage smallholder empowerment. A possible way to temper these problems is to restrict the private sector to those stages of a project which lend themselves to commercial operation.

Cooperatives: Cooperatives should in principle be the ideal institution for integrating all stages of the production process and ensuring farmers are fully represented. However the experience of cooperatives, mainly due to poor farmer representation, bureaucratic inefficiency and political manipulation has not been very good.

Voluntary organisations and NGOs: These have so far not played a major role in contract farming.

Crops Characteristics for Contract Farming

Coffee and tea are typical shrub crops. They differ from tree crops in that they need a high input of manual labour for pruning, weeding and harvesting and the end product requires early processing and can be transported at low costs. Tea and coffee cropping are determined mainly by soil and climatic conditions. Cacao, rubber and oil-palms are tree crops. They have low inputs of labour for maintenance but long gestation cycles until they can be harvested. Until the 20th Century, cultivation of perennial crops for export took place mainly in estates or plantations because they are suited to economies of scale. Almost all types are now grown in plantations as well as smallholdings although some of the crops are better suited for large scale production. Some major examples of smallholdings in Africa and Asia are: Cocoa in Ghana and Nigeria; rubber in Malaysia, Nigeria and Sri Lanka; coffee on the Ivory coast, Kenya and Madagascar; oil-palm in West Africa and tea in Kenya and Malawi.

Coffee: In most cases production started on estates and then spreads to smallholdings. Because coffee growing is good on a small scale much of the expansion since 1930s has been in outgrower systems, especially in Colombia and East Africa. In Kenya, half the coffee is produced by the small farm sector. Estate production has recently gained in competitiveness because of the rapid application of technical advances like improved varieties, mineral fertilising and plant protection. However, coffee growing is adaptable to big differences in farming intensity and to various forms of mixed cropping and intercropping; it is not very susceptible to disease; its most important input is manual labour but no pronounced peaks in labour demand; and it can be dried on farm or processed in small pulperies.

Tea: Tea can grow on acid soils in cool moist areas poorly suited to other crops. Tea has therefore been used to bring development to poor areas. It has a productive life of over 100 years and levels of productivity that can be easily influenced by early management, although bad early management is extremely difficult to rectify. Tea also requires harvesting many times per year and the main labour requirement is for the actual plucking of the crop.

Smallholdings are difficult to organise because tea has to be plucked and processed within a few hours and because of the importance of improved techniques. Further, economies of scale are pronounced and there is a tendency towards large factories. Estates are quicker in the adoption of innovations and produce double the amount of smallholders. However, the welfare effect of smallholder tea production is obviously higher and has been especially successful in Kenya, Tanzania, Uganda and Rwanda.

Cocoa: Cocoa, unlike tea, does not have large economies of scale; processing facilities can be designed to deal with different sizes of yield. It is therefore produced in many different sized holdings. Smallholders predominate in West Africa. The size of cocoa plantations is usually a function of the labour capacity of the household and their ability to invest time and labour into a crop which does not produce until 6-7 years. Cocoa is also often grown by men whilst women continue to be in charge of subsistence farming.

Rubber: Rubber is grown by both large and small producers. It has a long gestation period; about 7 years from planting to production with traditional varieties. Manual labour is a main item of cost, although there is no pronounced peak in labour demand as tapping can be done on a regular basis. The costs of establishing rubber plantations are high although it is easy to introduce innovations, and economies of scale are pronounced as higher yields do not lead to higher labour demands. For smallholders, rubber tapping is often part of an integrated farming system as a supplementary source of income. Smallholders are situated mainly in Sri Lanka, Sumatra, Malaysia and Nigeria. Tapping can be done whenever there is labour available. Both smallholders and estates have their comparative advantages. Smallholders may not receive maximum yields because of irregular tapping, lack of manuring and poor cleaning. However, estates may be hit by falling prices whereas smallholders can wait and let trees rejuvenate.

Oil Palm: Oil palm was part of subsistence based production systems as food, cash, medicine, construction, soap, and fuel. A genetic breakthrough changed palm oil from a low yielding but integrated crop into a high yielding commercialised crop. Modern oil palm production lends itself to estates because of its positive response to weed and bush control, regular employment of labour force and the need to process soon after harvesting. The initial capital investment is higher than with other tree crops but lower than for tea and coffee and the gross returns are higher than other tree crops. In outgrower schemes, smallholder oil palm production no longer has the close relations with other farm activities that used to exist with traditional palm groves. Oil palm is now often grown in a 'dualistic system'; a small plantation separate from the other crops.

Examples of Crop Characteristics and how they Influence Contract Farming

Type of processing

The type of processing which a crop needs has a clear impact on the contracts and production relations under which it is grown. When a crop (tea and palm oil) needs quick processing, outgrowers are particularly vulnerable to the terms stated by the contractor if they have no alternative place to bring the crop. The fixed capital assets of palm oil processing plants represent a high proportion of the cost of processing, so the profitability of the firm depends highly on the ability to operate very close to plant capacity. The form the contract takes will therefore be influenced by the need to maintain maximum production. Many palm oil estates are government run for this reason and operate through the nucleus estate-outgrower system so the contractor is not wholly reliant on the outgrowers.

Importance of quality

The importance of quality in the produce will have a bearing on production procedures. In some cases this will be through the transfer of technology to the outgrower and an investment in their ability to farm. This is the case with tea in Kenya, for example, where proper husbandry of the plant at an early stage is critical for later productivity. Extension agents visit outgrowers and transfer some of the skill for plant establishment. In other cases, the company will specify uniform production procedures which are not easily transferable and which do not build on the outgrowers indigenous knowledge of their micro-environment. In other cases there is very little technology transfer. For example, rubber and oil palm are homogenous and do not justify an expensive system to make sure farmers follow stringent practices and stringent grading requirements. Processors of such products have relatively weak incentives to provide producers with new technology.

Labour intensity and frequency

The labour intensity of the crop, and the frequency of labour input needed, has an enormous effect on how it is produced. Tea has been heralded as the perfect contract crop because it is labour intensive throughout the year and yields a monthly income. Rubber is a promising smallholder crop for the opposite reasons, it is not very labour intensive and can be harvested whenever there is labour available. The effect of labour intensity and frequency is therefore likely to depend on what role the contract crop has in the larger farming and social system.

Regularity of income and investment needed

Rubber, oil palm and tea are attractive for smallholders because they receive a monthly cash income compared to the annual payments received for cocoa and coffee. 'An oil palm grower is like a civil servant - a regular income and no work'. However most perennials, in particular cocoa, rubber and oil palm take a long time to mature. Smallholders therefore have to introduce these into their cropping systems gradually, or accept credit, and the associated risks, until their crops produce.

Shifting costs

The shifting costs associated with different crops, are the costs to a producer of shifting to a new buyer. These have been identified as the principal determinant of agricultural marketing structures and the credit systems associated with them. If the costs were low, producers would be free to choose their buyers. However, buyers would not be able to extend credit because they would have no assurance that this would be paid back or that they would reap the benefits of technical assistance to growers. If shifting costs were high they could develop a permanent producerbuyer relationship but there would be no assurance that the benefits would not be appropriated by the buyers. The ease with which outgrowers can shift is heavily influenced by the ease with which product quality can be assessed, the perishability of the product and the need for purchased inputs.

Financial and Economic Viability of Contract Farming

One of the major issues of contract farming is how efficient it is in comparison to plantation production and whether it is financially viable to promote as a route to development. Some of the evidence on this issue will be considered, by analysing the capital investment structure and the arguments made for smallholder efficiency.

Capital Investment

Contract farming usually involves a variety of actors including international agencies, government ministries or parastatals and private firms. In some cases the role these actors play is so intertwined that it is difficult to classify the scheme in any sector. This raises some important questions in assessing the financial viability and sustainability of smallholder schemes. How is the risk shared within these projects? For example, sometimes producers receive public credit at a subsidised rate. Does contracting with local producers allow foreign firms to shift the burden of financing crop production onto local governments and international aid agencies and benefit from indirect subsidised loans?

The capital investment levels per hectare and per farmer in contracting projects are high. This is especially so in settlement schemes where land and infrastructure have to be developed. Most projects revolve around high cost processing units. The major source of finance for smallholder projects are the government, local banks and companies and international funding agencies. The types of finance used include equity investment, loans, grants, services provided in kind and selfgenerated funds. Few smallholder projects have more than token equity investment (apart from smallholder land and labour investments) because of the riskiness of the venture from the viewpoint of the funding agency, and the readier availability of loan finance than equity capital.

Marketing and Price

Empty rhetoric to the contrary, contract farming represents the antithesis of free market forces. Contract farming schemes usually arise because of imperfections in the market environment that do not allow normal price signals to regulate supply. In many cases the market niche is either so narrowly defined or so unfamiliar that growers would not produce the crop unless they were under contract in a vertically integrated market system'. Thus, contract farming is in fact a response to market imperfections. The contracting scheme then tries to internalise the market for the commodity and avoid market competition.

A market monopsony is an essential component of contract farming as it is the only way to ensure that companies can secure a return on their money. This can be difficult with crops, such as coffee and cocoa, which are relatively easy to process and to market independently. It is usually when farmers begin to behave as free market actors, selling their goods to the highest bidder, that contract schemes run into problems. Many companies often depend on the state to ensure they receive a market niche that is kept free from too much competition.

If the costs to a producer of shifting to another buyer are low, producers would be free to choose their buyers and bargain over the price. However, the relationship between producers and buyers would be transitory and the buyers would be less willing to supply the producer with credit, inputs or technical assistance if they were not assured of a return on their investment. If shifting costs for the producer are high, then the contractor can be assured that they will have a return on their investment but the producer would not be able to bargain over price.

The shifting costs of a crop are highly dependent on crop characteristics such as the level of inputs required, the perishability and need for processing, and the ease with which quality can be assessed. Many producers do in fact have only one channel to which they can sell their produce. This is not necessarily bad for the producer because they have a guaranteed market for any amount of production of acceptable quality. However, it can leave the producer in a weak bargaining position on price and vulnerable to manipulation by project authorities.

The private traders however had a short term perspective and did not invest in stocking sufficient inputs, or supplying these to remote areas. The productivity of coffee fell during this period and farmers were hit hard when coffee prices slumped and private traders disappeared. In Thailand, contract farming has failed because the market is too competitive and buoyant. Growers and contractors both have multiple potential business partners, products and non-agricultural income sources. These do not meet the condition of quasi-monopsony which are crucial for contract farming. Contractors fear that growers will apply inputs supplied for contract crops to non-contract crops and contract defaults are the rule rather than the exception.

Market fluctuation presents a major problem for most of the commodities grown under contract. Many are grown in special market niches that are vulnerable to international market swings and boom and bust cycles. There has been a general decline in the terms of trade for plantation crops as compared to manufactured goods. Prices are also characterised by volatility, which are in part a consequence of the difficulty of adjusting to price changes. Smallholders are likely to respond to price swings by shifting their resources into another area of production. Contractors often try to forbid this diversification but for smallholders in a volatile market it may be essential for survival.

Price Formulas

The main factor in encouraging smallholders to join contract farming projects is the price the project authorities will pay for the product and a guaranteed market. This has widely been accepted as one of the best guarantors of success in a contract farming scheme. There is plenty of empirical evidence that smallholders will drop production of a crop, if at all possible, when the price falls. There are also instances where smallholders will continue to produce, despite conflicts with the management and difficult conditions, if the price paid for their crop is good, the market guaranteed and the payment prompt. Much of the success for smallholder projects depends on whether the contractor is willing, or able to meet these requirements. A lot therefore also depends on what alternative markets are available to the smallholders and the nature of their dependency on the producer.

Companies and governments have therefore tried to counter the volatility of the market and find ways to stabilise the price. A good price formula can help in sharing the costs and benefits between producer and processor. Table 3 illustrates five of the price formula which are used in contract farming. In a number of countries, national price stabilisation funds for particular projects have been established. For example, there is a rubber fund on the Ivory Coast, and a scheme for oil palm, copra and coffee in Papua New Guinea. Malawi has four smallholder crop authorities for tea, coffee, sugar and tobacco. These do not have individual price stabilisation schemes, but at the national level the losses of one authority tend to be balanced by the profits of another. In the absence of such schemes, farmers diversify their production base.

Access to Credit

Access to credit is one of the big incentives for smallholders in joining contract farming schemes. The credit can be given in cash, in kind, or in the advance of services or capital inputs. Loans are usually given on the security of the land or the anticipated value of the export crop. Loan recoveries are usually made from crop sales or as service charges. Sometimes the farmers obtain loans separately from an existing national credit agency or a bank, in which case the contract itself can serve as collateral. Glover has found that transnationals are increasingly reluctant to finance contract growers and have attempted to persuade state banks to assume this responsibility.

The advance of credit is a vital part of contract farming and often the only way the smallholder can enter the market. Where tree crops are cultivated, whatever the scale of operation, one function of capital is to tide the producer over the period of maturation. Debt is also more likely to be a problem for tree crops which have a long maturation period.

Another major problem mentioned with credit is that smallholders can be locked into a deteriorating debt situation. A grower may enter a contracting relationship and then be unable to terminate it if the company or project deducts payments and the expected benefits do not materialise. In such a situation the grower may have to stick to the same company and is effectively at their managerial mercy. The deteriorating debt situation can be exacerbated because of favourable terms offered to growers in the early stages of a project. The contractor will offer favourable terms on credit and price in order to attract competent suppliers, but may be unable or unwilling to sustain these in the long term.

Welfare and Equity Effects of Contract Farming

Contract farming schemes need to be assessed in a national and regional context as their effects on non-scheme farmers, towns and villages are important. What is favourable for the scheme and its participants may harm other actors and sectors of the regional economy. They can have the effect of concentrating public resources on small groups of farmers. Contract farming is often favourably compared to plantations because they do not create an 'enclave' effect but the evidence on this is far from conclusive. Further, participation in contracting schemes is usually one component of a growers diversification strategy and has to be evaluated as part of the whole livelihood and farming system.

Contract Farming Exclude Small Farmers

Many critics of contract farming allege that it relies on large farmers. This is a generalisation. However, companies do often prefer large farmers because it cuts down on transaction costs and allows for a more uniform quality of product. A large number of smallholders can be costly and time consuming to organise. Some contract farmers are simply legal fronts for large companies who are trying to avoid prohibitions of land ownership. Contract farming, particularly of perennial tree crops, is also very attractive for absentee landlords as it requires a minimum of supervision and labour. Some contract farming schemes have not only excluded small farmers but had the effect of evicting tenants because these landlords have decided to plant their land with tree crops.

Glover finds that when companies deal with small farmers there are usually several factors present:

— The area most suitable for production is characterised by smallholders so the companies just work with whoever is available.

— The government may encourage firms to use smallholders. Companies will be especially willing to do so if the inclusion of smallholders provides the companies with access to credit at concessional rates from financial institutions. Smallholders also have less power to influence the government or welfare of smallholders, has to be judged from a long-term perspective. A fluctuating income can be very risky for settlers who can be thrown off their leased land if they are not able to pay for services charged to them by the contractor, or to keep up with their debt repayments. The welfare effect of increased income therefore has to be balanced against continuing debt of outgrowers and the frequent need to purchase food.

Equity Amongst Smallholders

Contract farming not only tends to exclude smaller farmers but there is considerable differentiation amongst smallholders under contract. The difference is often between the rich capitalist farmer who employs labour and substantial middle peasantry who may earn an investible surplus but depend largely on household labour. Small farmers will use a larger proportion of their income in basic needs, whilst larger growers can invest in productivity enhancing technology. 'Nevertheless contract farming appears to be the most slowly differentiating route yet known for the transition from traditional local market agriculture to highly commercial capitalist agriculture'. In many instances this process has been helped by a concerted government effort. Central control has helped to restrain the emergence of economic inequality amongst smallholders by preventing opportunities for the wealthier or more progressive farmers to gain control over the production procedure or to reinvest their profits locally.

Regional Development

At the national and regional level, public support for contract farming appears to have led to the concentration of public resources. On occasions, smallholders are actually displaced from their land in order to make room create legal problems in case of a contra dispute.

— Smallholders may have lower costs of produ because they rely on family labour and are w to take more risks. If they are given acce technical advice, smallholders can freque provide high quality, labour intensive crops.

Contract farming usually requires a legal title to l which already excludes the landless and tenants. Man the smallholders who become contractors already h some experience of cash cropping and some n agricultural income alternatives which allow them to b the risk of signing a contract. Further, many contr farming schemes are located in high potential areas w fertile soil and access to irrigation and transport.

Most contract farming schemes have a fairly ste criteria for inclusion. Apart from secure title to land, the can include a minimum land size, good health, a prove ability to hire labour or enough family labour, man projects prefer a married status and sometimes eve education or prior experience with the crop. Inclusion o a project, especially successful ones where the waiting lis is long, can depend on having the right leverage o political patronage. The use of this frequently enables absentee landlords to join schemes which are designed specifically to assist smallholders. Many small farmers are accepted initially and then required to undergo training before becoming full time members.

Welfare of Smallholders

Most contract farming projects do appear to contribute to smallholder welfare and improve farmer incomes, at least in the short term. The risk of contract farming and the for settlement projects. The evidence of the linkages created by smallholder programs is mixed. The evidence on employment creation from smallholder production is mixed and crop specific.

Contract farming schemes do generate much employment in part because the crops are labour intensive and the hiring of labour by contract growers to deal with seasonal work is common. Most of this employment is on the farm and some is in support services like processing and transport. It is generally agreed that when smallholders do employ labour they pay amongst the poorest rates.

Collective Action and Farmer Organisations

Considerable attention has been paid to whether or not contract farming encourages collective action amongst smallholders and whether this can improve conditions and lead to more general local development. Contract farming does have some elements which it seems would encourage collective action; many smallholders, a common adversary, a degree of interdependence and the contract, a tangible and common issue for negotiation.

Many of the instances mentioned of outgrower resistance and manipulation of the terms of the contract are not collective but individual acts. In fact some of these acts, such as the use of personal and patronage ties to have produce upgraded, actually work against collective action. Others are simply individual ways to gain a larger share that would not work collectively, for example, adding stones to increase the weight of a produce or selling the produce on the open market. The record for collective action with contract farming is mixed although it does appear to hold potential for the development of farmers organisations.

Collective Response from Contract Farmers

Heterogeneity among growers either in ethnic or economic composition. For example, outgrowers can vary between absentee landlords and full-time smallholders. Absentee landlords may turn their farms over to the contracting company and be quite content to escape land expropriation under agrarian land laws. They may not be particularly concerned about the product price and have little in common with fully profit-oriented smallholders.

Another conflict may be one between simple producers and ones who have access to multiple markets. The latter can take advantage of the simple producers and may have no interest in reforms designed to improve the market access of simple producers. They also have no interest in stabilising prices because they derive their margins from these fluctuations. Although the interests of the larger growers may diverge from those of the simple producers, it is very difficult to exclude them from growers associations.

The contracting relationship often encourages the cultivation of individual self-interest even if growers do understand there may be long term benefits of acting collectively. No contract is perfectly contingent, and many offer individual incentives to growers which encourages them to cultivate cordial relations with the management.

Encouraging Collective Action

Homogeneity, both in ethnic background and in economic status, encourages collective action. Sometimes, when there are many smallholders, the company will itself organise some form of organisation or cooperative to coordinate production. Whether or not this group restricts itself to a simple coordination function or starts to lobby for better conditions depends on the extent to which the company is perceived as a common adversary and the relative dependencies of the growers on the company.

In Ghana, farmers were evicted from their land to enable the establishment of a large oil palm nucleus estate. Some of the land, in relatively equal parcels, was then given back to the farmers who became contracted outgrowers. The lack of collective action can, in part, be explained by the heterogeneity of the contracted workers. Contract farming, because it involves negotiation, does appear to present an ideal opportunity for the emergence of farmers organisations.

9

Distribution in Contract Farming

In principle, contract farming could be an institutional arrangement that enables farmers to access markets. While contractual arrangements can vary by crop and by country, contracting is a form of joint production where the grower supplies tools, land, labour and management while the processor supplies technical assistance, some inputs such as seeds or pesticides and undertakes to buy the grower's output at a pre-determined price.

From the point of view of the processor, this arrangement ensures raw material supplies of the desired quality. From the point of view of the grower, such an arrangement provides an assured market and hence reliable income. Without a contract, risks would be too much and few small growers would want to produce these crops. For this reason, Glover described contract farming as an institutional arrangement that combined the advantages of plantations and of smallholder production.

These theoretical benefits, notwithstanding, contract farming has been controversial and has been criticised for being exploitative. Between the giant corporation and the small farmer, bargaining power surely lies with the former. Also, in practice, growers have encountered problems with respect to manipulation of quality standards, poor technical assistance, and sometimes plain cheating and deliberate default. As a result, Glover concluded that research must "systematically examine successes and failures and from then draw generalisations about the conditions under which CF (contract farming) can operate profitably and to the benefit of small farmers".

Poultry Industry in Andhra Pradesh

The poultry industry has seen rapid growth in India. Between 1980/81 and 1998/99, poultry meat production increased about 3 times from 250,000 tons to about 770,000 tons. Correspondingly, its share of total meat output rose from 20% to 27%. During the same period, egg production increased from 10,000 to 29,000 million. Andhra Pradesh is the leading poultry meat producing state within India. It accounts for over one-fifth of poultry meat as well as egg production in the country. The growth of poultry industry in Andhra Pradesh has been even more remarkable than the national growth rates. For the period 1980/81 to 1998/99, poultry meat production increased by 4.5 times while egg production rose by 3.5 times. About 30% of its broiler output and 15% of egg output are exported to other states within India.

The impressive growth in the poultry meat industry is the result of technological breakthroughs in breeding, feeding and health, and sizeable investments from the private sector. The expansion in supply has been spurred by rising incomes and has resulted in lower poultry prices in south India where much of the growth has occurred. The poultry sector is, however, highly prone to production and market risks, which periodically affect the profitability of broiler production particularly on the small farms.

These risks also threaten the profitability of the industry engaged in breeding of chicks and manufacturing of feed, vaccines and medicines. In order to minimise the risks to the producers and sustain the profitability of the industry, some large poultry firms began integrating their activities with that of broiler production through the institution of contract farming as early as in late 1980s. A largescale integrated operation would typically include the raising of grandparent and parent flocks, rearing of day-old-chicks, feed milling, provision of veterinary services and contract production. Such integrators are most common in southern and western regions of the country.

Contracting in Poultry Production

In a poultry contract, hatcheries provide day-old chicks, feed and medicines to contract growers. The contract growers supply land, labour and other variable inputs. At the end of the production cycle, the farmer receives a net price that is pegged to an industry price set by a group of hatcheries. The industry price fluctuates within a narrow range and is a lot more stable than the retail price. Thus, the farmer receives considerable price insurance.

For sharp upward deviations of the retail price from the industry price, farmers receive an incentive. This practice presumably lessens the incentives to default on the part of growers and reflects the competition from the non-contract sector. The farmer is insured for mortality rates up to 5%. Beyond that the farmer bears the risk of loss. This controls moral hazard and provides incentives for farmers to supply their best effort.

A company representative who sorts out problems especially regarding disease visits the farmer daily. According to company accounts, the processor spends time and resources in screening producers for reputation and prior experience. The broiler contract is an instance of a "production management" contract where the processor supplies inputs and extension, advances credit (in kind), provides price insurance and monitors grower effort through frequent inspections. The detailed monitoring is because of the considerable credit advanced by the processor that provides more than 90% of the cost of production in terms of the value of inputs. Because the frequent monitoring controls for moral hazard, it is also conducive to insurance. The frequency of contact also would mean that the processor incurs considerable transactions costs.

Data from Field Survey

The data was collected from a primary field survey of contract and non-contract producers. The survey was undertaken in the year 2002-03 to collect the required information for the year 2001-02. The sample producers were interviewed to collect the required data, using pre-tested questionnaires, specifically prepared for each case study. Survey data was based on recall memory of the households but it was also supplemented with the records maintained by both contract and non-contract producers.

Detailed information was collected about the socio-economic characteristics of the sample farmers, production-portfolio, item-wise and cycle-wise cost of broiler production, yield levels, labour use, and cost of marketing and acquiring information for various activities. A sample of 25 contract producers and an equal number of non-contract producers were randomly selected from 10 villages of Rangareddy, Mehboobnagar and Nalagonda districts in Andhra Pradesh. A majority of the contract farmers were associated with a leading poultry integrator.

The survey instrument consisted of four parts. In the first part, information about village level infrastructure was collected. This consisted of distance from various infrastructure facilities such as roads, railways, telephone, post office, regional rural bank, animal feed shop among others. Table 1 compares the availability of infrastructure across contract and noncontract farmers. The big difference between contract and noncontract farmers is the better access of noncontract farmers to credit facilities whether it is cooperative credit society or the regional rural bank or the primary dairy cooperative society.

Table 1. Access to Infrastructure

Item	*Noncontract*	*Contract*
Distance to urban area	28.36	17.16
Distance to coop credit society	0.43	2.48
Distance to regional rural bank	1.2	6.84
Distance to primary dairy cooperative society	0.48	8.5
Distance to vet. Hospital	.8	0.71

The second part of the survey elicited information about the farmer including age, schooling, experience in broiler farming and previous occupation. Table 2 summarises the differences between contract and noncontract farmers in terms of individual characteristics. Notice that the sample of noncontract farmers is twice as experienced, slightly more educated and yet a little younger than contract farmers. The sample of noncontract farmers also contains a substantially higher proportion of farmers who are specialised in poultry farming. The noncontract farmers are also those who have migrated from agriculture related occupations compared to contract farmers where the proportion of farmers with non-agriculture related occupations is large. Examples of such previous occupations include in sectors such as pharmaceuticals, electrical hardware, cement, police, clothes and wine retailing.

Table 2. Characteristics of Poultry Producers

Item	*Noncontract*	*Contract*
Experience in poultry	9.8	4.9
Age	36	39
Years of schooling	11.6	10.9
Proportion of farmers whose main occupation is poultry	72	36
Proportion of farmers whose subsidiary occupation is poultry	28	64
Proportion of farmers whose earlier occupation was in agriculture/poultry/ dairy/ agricultural labour/ agriculture-related business	75	58
Proportion of farmers whose earlier occupation was in non-agriculture	25	42

The third part of the survey collected information about the inputs, outputs and revenues from the last 6 production cycles of each grower. Table 3 presents information about the levels of input use per production cycle for contract and noncontract farmers. Note that the numbers are averaged twice - first over production cycles for each grower and then across all growers within the class of contract and noncontract farmers. The production process in poultry consists of transforming baby chicks into fully-grown birds. Besides chicks, the inputs into this process are feed, medicine, labour and time. Production is not instantaneous. For this reason, productivity measures have to be normalised with respect to time as well. With respect to capital assets, contract farmers have more equipment as measured by brooders, feeders and water facilities but operate in a smaller area.

Table 3. Input Use by Poultry Producers: Averages Per Production Cycle

Item	*Noncontract*	*Contract*
Time: Cycle length (days)	48.4	42.6
Litter, quantity (quintals)	13.2	17.8
Litter , value (Rs.)	991	1,375
No of chicks	6,891	8,149
Chicks, value (Rs.)	70,217	96,558
Feed qty (quintals)	276	277
Medicine + Feed, Value (Rs.)	244,615	307,246
Vaccinations per cycle	2.53	3
Vaccination, cost	6,174	8,148
Veterenian fees (Rs.)	268	566
Labour cost (Rs.)	5,076	6,152
Family labour (male) (Days)	26	23
Family Labour (female) (Days)	4	3.2
Hired labour (male) (Days)	76	104
Hired labour (female) (Days)	37	33
Electricity, value (Rs.)	2,930	2,925
Bulbs, (Rs.)	335	274
Disinfectants (Rs.)	861	993
Total shed area (square feet) per grower	9501	6335
No. of Brooders per grower	12	24
No. of Feeders per grower	158	175
No. of borewells per grower	0.36	0.6
No. of overhead tanks per grower	0.72	0.68

From table 3, it can be seen that noncontract farmers have longer production cycles, lower flock sizes and correspondingly spend less on medicine, feed, vaccination, veterinarian fees and labour cost. Noncontract farmers depend much more on family labour than contract farmers. Medicines and feed is the single largest item of variable cost. Interestingly, even though noncontract growers manage lower flock sizes, they use almost as much quantity of feed as the contract growers.

Table 4 compares the outputs and revenues (from bird sales) of contract and noncontract producers across all production cycles. As contract producers have larger flock sizes, their output is also larger whether measured by the number of birds or the total weight of birds sold. However, the average weight of a bird is pretty much the same across contract and noncontract growers. The average revenues per kg of bird are much lower for contract farmers reflecting the netting out of input costs by the processor.

Table 4. Output and Revenues: Averages Per Production Cycle

	Noncontract	*Contract*
Output: # of birds	6583	7808
Mortality: # of birds	302	388
Average total weight of birds sold (Kgs)	12105	13638
Average Weight per bird, Kgs	1.87	1.87
Revenues from bird sales (Rs.)	355,732	37,217
Average Revenues/Kg of bird sold	29.1	2.62

The final section of the survey is relevant only to contract farmers and it obtained information about the contract between the producer and the processor. In particular, this section contains information about the nature of input sharing between the producer and the processor. As noted in the earlier section, processors supply chicks, medicine, feed and veterinary services. Growers supply land, buildings, labour, and other variable inputs such as electricity and disinfectants. Using the information on input sharing, Table 5 computes the total value of variable inputs and the value of inputs supplied by the grower. For the farmers not on contract, the two figures are the same. But this is not so for contract growers. For them, the processor supplies most of the inputs measured in value terms. On average, the out of pocket expenses for inputs for contract farmers is less than 3% of total input costs.

Table 5. Input Sharing: Averages per Production Cycle

Value of all inputs (Rs.)	331,468	424,200
Value of inputs supplied by farmer (Rs.)	331,468	12,249

Efficiency of Contract Production System

As a processor has the choice of contracting with growers or procuring the bird from noncontracted growers, contract production should reduce processor's costs if it is to be observed. The efficiency of contract production is therefore evaluated relative to the production costs of independent growers. Costs in contract production could be lower than in noncontract production in two distinct ways. First, because of technology and management practices brought by the processor, contract production could be more efficient than noncontract production. Second, if the processor can access some inputs such as insurance and credit at lower cost than growers, then contract production could be cheaper than noncontract production even if production efficiency is unchanged.

Interest arbitrage possibilities exist even in the absence of contract production and such gains are realised by the creditors of noncontract growers. Therefore, from an economy wide point of view, it is the cost reduction from better technology and production practices that should be counted as efficiency gain. Table 6 displays the major components of production costs in both modes of production. It can be seen that the cost structure is comparable across the two production systems. Feed, medicine and veterinary services accounts for about 75% of total variable cost. The expenditure on chicks is about 20-22% of cost while other variable costs such as labour and electricity constitute only 3% of total costs.

Table 6. Cost structure of Poultry Production: Averages per Production Cycle

	Non Contract	Contract
Chicks, value (Rs.)	70,217 (20%)	96,558 (22.5%)
Feed & Medicines (including vaccinations and veterinarian fees)	251,058 (77%)	315,959 (74.5%)
Labour, electricity & other inputs	9,203 (3%)	10,344 (3%)
Total	331,468	424,200

As the cost of poultry production is primarily the cost of chicks and feed, the technology is characterised by constant costs. Table 7 displays the results of regressing the total variable cost on total output (measured in kgs). Recall the data set consists of observations from up to 6 production cycles for 25 contract and 25 noncontract growers. Thus, the error term will contain a producer-specific component. To take that into account, all standard errors are corrected for heteroskedasticity as well as dependence stemming from the correlation of errors from the production cycle of a particular producer. The regression is done separately for contract and non-contract producers. The predicted value from these regressions is graphed against the dependent variable in Figures 1 and 2.

Table 7. Cost function of Poultry Producers

Dependent Variable: Variable cost per production cycle

Variable	*Non-Contract*	*Contract*
Constant	6635 (2.17)	-11192 (-1.45)
Output (kgs)	26.28 (110)	30 (36)
R^2	0.983	0.98
# Observations	145	140

t-statistics in parentheses.

From the results, it can be seen that it costs Rs. 30 to produce a kg of bird under contract production and Rs. 26.22 under non-contract production. This suggests that processors would be better off by abandoning contract production and should instead buy from noncontract growers. However, because contracting is a form of joint production, it should be remembered that it is the processors who determine the feed, medicine and chick costs of contract growers. Therefore, these numbers are not necessarily indicative of competitive prices but may well be a sign of transfer pricing. To have cost figures that reflect competitive prices for feed and medicine, we can recalculate contract production costs using the price paid by non-contract growers. This price is recovered from cost data of non-contract growers. Their unit feed costs depends on feed prices as well as the feed-conversion ratio.

The feed price paid by non-contract growers can therefore be obtained as the ratio of their feed cost and their feed-conversion ratio. The above procedure assumes unit costs for feed and medicines are constant. This is indeed the case as the feed and medicine costs are proportional to output. For noncontract growers, feed and medicine costs work out to Rs. 19 per kg of bird, which is calculated as the slope coefficient in a linear regression of feed and medicine costs on output.

In the second step, we regress feed quantity on output and obtain the feed-conversion ratios as 1.88 and 2.15 for contract and noncontract growers. In other words, noncontract growers use 2.15 kgs of feed to produce a kg of bird. Thus, the price of a kg of feed and medicine works out to be (19/2.15) = Rs. 8.84 where we are assuming medicine requirements (medicine, vaccines and veterinary services) is proportional to feed.

In the third step, we use this price and the feed quantity used by contract growers to recalculate their feed and medicine costs that would obtain if they were charged the same prices for feed and medicine as noncontract growers. To these costs, we add the observed costs for chicks and other inputs that are incurred by contract growers and we thus obtain a simulated figure for total costs for each production cycle of the contract grower. Finally, by regressing the simulated costs on output, we obtain the marginal (and average) costs for the contract grower as Rs. 24.3. Compared to the marginal costs for the noncontract grower of Rs. 26.2 per kg, contract production involves a saving (relative to procurement from noncontract growers) of Rs 1.9 for every kg of bird.

As contract production is more efficient than noncontract production at zero rate of interest, taking interest costs into account can only increase its relative advantage. Rs. 1.9 is a minimum bound on the surplus from contracting relative to contract production. For instance, assuming an annual interest rate of 15% per annum for both modes of production, the marginal costs of noncontract production rises to Rs. 26.77 while that of contract production rises to Rs. 24.7. The savings from contract production therefore rises to Rs. 2.07 per kg of bird per production cycle.

Table 8-Feed-Conversion Ratios

Dependent Variable:	*Feed Quantity*
Constant	7.98 (1.8)
Output (kgs)	2.16 (81)
Output x Contract Dummy	.029 (-3.9)
R^2	0.983
# Observations	145

t-statistics in parentheses

It might be argued that the borrowing cost of funds for growers would be higher than for processors that have access to institutional finance. For instance, if noncontract growers can borrow only at 21% when processors can borrow at 15%, the cost of noncontract production goes up to Rs. 26.95 while that of contract production does not change. It would seem then that the surplus from contract production rises to Rs. 2.28 per kg. However, note that Rs. 0.21 of this surplus is because of interest arbitrage by the processor and not because of greater production efficiency in contracting. The higher efficiency of contract production is driven by its lower feedconversion ratio.

The coefficient on the interaction variable estimates the difference in feedconversion ratios between the two groups of producers. The results are presented in Table 8.

Once again the standard errors are corrected for heteroscedasticity as well as correlations between errors stemming from producer specific components. It is clear that the difference between the feed-conversion ratios for contract and noncontract production is statistically significant.

Cost to Growers and Processors

The value of home consumption, if any, is also imputed to revenues. Inputs consist of chicks, feed, medicine, vaccine, litter, veterinary fees, labour, electricity and disinfectants. For contact growers, however, the processor advances most of the value of inputs. Compared to the noncontract grower, the contract grower needs less working capital and therefore incurs lower interest costs. Information on the opportunity cost of funds for contract and noncontract farmers is, however, missing in the survey.

From studies of rural finance, we know that informal credit is widely prevalent and that it is more costly than credit from institutional sources. According to the all India rural credit survey, formal sector accounted for 53% of all rural credit in 1991. Moneylenders and friends or relatives account for the rest. More recent data from the World Bank indicates that access to formal sector credit is very limited for poorer households. According to the same survey, the median interest from banks in 2003 was 12.5% per annum while the average interest rate from informal sources was 48%.

For credit from institutional sources, transactions costs are also significant. These arise because of distance to financial institutions, cumbersome procedures and bribes ranging from 10% to 20% of loan amount. As a result, the effective cost of credit from formal sources is likely to be greater than the median interest rate. A survey in 2001 of the poultry sector reports that interest rates on commercial loans were typically around 15% per annum. As informal credit is more costly than this, an interest cost of 15% per annum can be taken to be a lower bound to the cost of credit for growers.

Table 8 compares the incomes from poultry farming for interest rates ranging from 15% to 30% per annum. As one would expect, the returns to noncontract growers declines significantly as interest rates rise while the contract farmers are almost completely insulated from credit costs. The returns are equal for both modes of production at a 10% rate of interest. For interest rates higher than 10%, the returns for contract growers are higher than that of contract growers. If we take 15% to be a representative borrowing rate for growers, contract farmers earn on average Rs. 0.15 per kg more than noncontract farmers, i.e., about 7% more than the per kg average earnings of a noncontract grower.

If, in fact, 15% is the also cost of funds for processors, then contract production yields a surplus of Rs 2.07 per kg out of which farmers receive Rs 0.15 or about 7% of surplus. Note that the entire remainder is not the profit of processor as the processor also incurs costs in administering and managing contracts that are not taken into account here. While the gains to contract farmer are not trivial in magnitude, they are not statistically significant when interest costs are 15% per annum. The standard errors of the difference in returns between contract and noncontract growers are corrected for heteroskedasticity and within cluster correlation. As the relative advantage of contract farmers increases with interest cost, the statistical significance of the difference also increases.

Correcting for self-selection

We can compared the average returns of contract farmers with the average returns of non-contract farmers. While this is useful to demonstrate the distribution of surplus from contracting, it is a biased measure of the gains that actually accrues to contract farmers because it does not take account of the fact that the processor purposively selects the contract farmers. Hence it is likely that the population of contract farmers differ from the population of non-contract farmers.

To take this into account, we adopt the treatment effects models from the program evaluation literature. Let y_{li} be the returns with contracting and y_{i0} the returns without contracting for the ith producer. Clearly, only one outcome is observed for each producer. The average treatment effect is) (E $(y_{li} - y_{i0})$ while the treatment effect on the treated is E $(y_{li} - y_{i0}) \mid C_l = 1$. Comparing mean outcomes of contract and noncontract farmers are equivalent to E $(y_{li} \mid C_l = 1 -$ E $(y_{l0} \mid C_l = 0)$ which is in general not equal to either the average treatment effect or the treatment effect on the treated.

In a regression framework, the treatment effects model is given by

$$R_t = a + bC_l + c'X_t + \varepsilon_1$$

where R_i is the net returns of the ith producer, C_i is a dummy variable that takes the value 1 if farmer *i* is in contracting and takes the value 0 otherwise. X_i is a vector of control variables such as farmer characteristics (experience, and e's are zero mean random variables. *b* measures the impact of contracting on mean returns. Under the assumption of homogenous treatment effects, *b* identifies the average treatment effect as well as the treatment effect on the treated.

If the variables in the X matrix include all the variables that influence whether a producer is a contract grower and if these variables are not correlated with the error term, then ordinary least squares estimates of (1) are consistent. These are displayed in the second column of Table 9 (assuming an interest rate of 15%). The estimate of the impact of contracting is comparable to the difference in group means (in Table 8) and is insignificant.

Standard errors are corrected for heteroskedasticity and within cluster correlation. The other variables have the expected signs and are significant. The other variables in the regression are experience, experience squared, season, season squared and value of assets. Season is a variable that takes values from 1 to 12 and identifies the month in which production begins. Thus a production cycle with a season code of 1 begins production in early January and the output enters the market after mid-February. The season variable is therefore meant to take account of the seasonality in prices and production. As the seasonal trend is quadratic, we have also included the squared term of season.

Table 9. Returns to Poultry Producers: Average Income Per Production Cycle (Rs/Kg)

Annual interest rate	Contract	Noncontract	Difference (t-value)
15%	2.20	2.05	0.15 (0.76)
20%	2.20	1.9	0.3 (1.49)
25%	2.19	1.66	0.44 (2.2)
30%	2.18	1.47	0.58 (2.84)

If treatment impacts are heterogenous, then additional controls in the form of interactions of the contract dummy with the demeaned explanatory variables must be included. These estimates are contained in the third column of Table 9. Note that under heterogenous effects, the average treatment effect and the treatment effect on the treatment no longer coincide and the coefficient on the contract dummy identifies the average treatment effect. The magnitude and significance of the average treatment effect improves but it is still not significant even at the 10% level.

Standard instrument variable procedures are used to correct for the bias from the endogeneity of right hand side variables. Consider a participation equation such as

$$C_i = y_1 + y_2 Z_i + u_t \tag{2}$$

where Z_i is a vector of variables that matter for participation. Variables in Z_i will overlap with variables in X_i. Identification requires that there be at least one variable in Z_i that is not in X_i. If this condition is met, the predicted value from (2), $\hat{C}_i$ can be used as an instrument for C_i in regression equation (1). This would yield a consistent estimate of b provided the instruments are uncorrelated with the error term in equation (1)

Table 10 displays the estimates from a participation equation. Growers who are at more distance from credit facilities, less specialised in poultry and with previous occupational backgrounds in non-agriculture is more likely to be contract growers. In addition, experience and schooling negatively affect the probability of being a contract grower. These results are consistent with anecdotal accounts in poultry of processors wishing to contract with growers with weak bargaining power. Such an outcome was discussed by Key and Runsten.

Table 10. OLS estimates of Treatment Effects Under Varying Interest Rates

Interest Rate	*Average Treatment Effect (t-value): Homogenous Effects*	*Average Treatment Effect (t-value): Heterogenous Effects*
15%	0.16 (0.97)	0.21 (1.39)
20%	0.3 (1.85)	0.35 (2.36)
25%	0.44 (2.69)	0.49 (3.28)
30%	0.58 (3.47)	0.63 (4.16)

Of the variables in the participation equation, distance from rural banks is the variable that seems to best fit the specifications of an instrument. It has predictive power in the participation equation and, conditional on the contract dummy, it is also redundant in the income equation.

Table 11. Probit Equation: Factors Influencing Participation in Contracting

Explanatory Variables	Coefficients (t-values)	Marginal Effects
Distance from Regional Rural Bank	0.17 (6.62)	.07
Years of Schooling	-0.06 (-2.2)	-0.02
Experience	-0.54 (5.94)	-0.21
Experience squared	0.02 (4.52)	0.009
Whether previous in Non-Agriculture	0.74 (3.17)	0.29
Whether Poultry is main occupation	-0.84 (-4.01)	-0.32
Constant	2.45 (5.6)	--
No. of Observations	50	50

The variables of schooling, specialisation in poultry and previous occupational background, could affect income in addition to their effect through participation. As it turns out, however, they are insignificant in the income equation possibly because their impact on the outcome works through grower's experience in poultry that is independently controlled for. Hence, these variables are also good instruments.

Table 11 contains estimates of equation (1) where the dummy for participation in contracting is instrumented by the predicted probabilities from the participation equation. Standard errors are corrected for heteroskedasticity and within cluster correlation. The other variables in the regression are experience, experience squared, season and season squared. The dependent variable in this regression is computed assuming the cost of credit to be 15% per annum.

Risk Shifting

Calculating the mean income gains from contracting provides only a partial picture of the change in utility for contracting producers. A fundamental feature of contract farming is the shifting of risk from producers to processors. In the broiler contract, much of the price risk is reduced by the use of the bro-mark (set by the processors) rather than the market price.

The most straightforward way to estimate risk shifting would be to compare the variability of net returns of contract growers with that of non-contract growers. But this comparison would once again be subject to bias because of the use of incorrect counterfactual. Knoeber and Thurman propose that the variability of net returns of contract growers be compared to the hypothetical or simulated returns that they would have received as "independent growers" i.e., if they had purchased inputs and sold their output at market prices and not contracted with the integrator.

Let σ_i denote the standard deviation for the ith producer. These are calculated from the data on 6 production cycles for each grower. Also let and denote mean standard deviation and mean coefficient of variation for the group of contract growers and noncontract growers respectively. They are estimated as the sample means of the σ's and v_i's and are reported in the first two rows of Table 12. The computations assume the lowest possible interest rate of 15% per annum. The table also reports the standard errors of these estimates. The figures show that the variability of returns of noncontract growers exceed that of contract growers by a factor of 8 or 10 depending on the measure of variability (standard deviation or coefficient of variation). However the estimate of average variability for the noncontract growers is not very precise because of the large differences in variability within the noncontract group. The coefficient of variation ranges between 0.23 and 4.3 for noncontract growers while it ranges between 0.023 and 0.26 for contract growers.

Table 12. Variability of Returns

	Noncontract	*Contract (Observed)*	*Contract (Simulated)*
Mean of Standard Deviations of Individual Growers	$\sigma_n = 29.2$	$\sigma_c = 0.26$	$S_C = 2.17$
(standard error)	(0.84)	(0.16)	(1.29)

There are two components of the simulation. First, for the inputs advanced by the processor (chicks, feed, medicine and vaccines), we value their cost using prices paid by noncontract farmers. Second, we use the price received by noncontract growers for their birds, bags and litter to value the output of these items by contract growers. As the prices received (for output) and prices paid (for inputs) by noncontract growers are not identically the same, we use the median figure in all the cases. In all imputations, we use figures from comparable production cycles. For instance, the price used to value a contract grower's output from production starting in January would be the median price of noncontract growers in the same month.

From the simulated series, we construct once again the mean and standard deviation of returns. Let si denote the standard deviation of the simulated series for the ith producer. Also let sc denote the mean standard deviation for the group of contract growers. This is reported in the last column of Table 12. The variability of the simulated series is of the same order of magnitude as the variability of returns for noncontract growers. On average, the standard deviation of the simulated series is more than 8 times greater than that of the actual series.

For each individual grower we compute the ratio of the standard deviation of the simulated series to the standard deviation of the observed series. For the 25 contract growers, the average of this ratio is 13.4. The median ratio is 8.25 and the distribution ranges from a minimum value of 2.7 to a maximum value of 91. At the median level, growers under contracting bear only 12% of the risk that would have been borne by them as noncontract farmers. In other words, 88% of the risk in poultry farming is shifted from farmers to processors as a result of contracting.

The statistical significance of the reduction in variability can be assessed for each grower by testing the hypothesis that the simulated variance for the ith contract grower equals the variance of the observed series. As the simulated and observed series are correlated, Knoeber and Thurman derive a Wald statistic that takes this correlation into account. The statistic is

$$T_i = (s_i^2 - \sigma_i^2) / [2 / n)(s_i^4 + \sigma_i^4 - 2\rho_i^2)]^{1/2}$$

where for the ith producer, s_i^2 and σ_i^2 are the sample variances of the simulated and actual series, ρ_i^2 is the covariance between the two series and *n* is the number of production cycles. Under the null hypothesis that the variances of the two series are identical, the Wald statistic is asymptotically standard normal.

The median value of the Wald ratio is 1.69, which means that for 50% of contract growers the null of no difference in variability is rejected in favor of the one-sided alternative that the variability is greater in the simulated series at the 5% significance level. The smallest Wald ratio is 1.41. Hence the null would be rejected in favor of the alternative for all growers at the 10% significant level. The reason that the differences are not statistically valid at the 5% level for some growers is because of the very small number of production cycles as a result of which the differences in variability are estimated imprecisely.

The risk reduction from contracting can also be assessed by testing the null hypothesis that the median value of i s and si are equal. This can be done by making use of nonparametric tests for difference in medians using paired data. The paired data in this instance involves the observed and simulated standard deviations for each grower. The sign test considers the number of times the difference between the simulated and observed standard deviations is positive. The null is rejected if the number of differences of one sign is too large or too small.

10

Commercialisation of Contract Farming

In contract farming, a central processing or exporting unit purchases the harvests of independent farmers. These purchases can supplement or substitute for company production. The terms of the purchase are arranged in advance through contracts, the exact nature of which can vary considerably from case to case. Contracts are generally signed at planting time and specify how much produce the company will buy and at what price. Often the company provides credit, inputs, farm machinery rentals, and technical advice. It always retains the right to reject substandard produce.

Contracting is most commonly practiced by food processing companies. Since their processing plants have high fixed costs, these companies have an interest in keeping raw material inflows at a steady level, close to plant capacity. Reliance on open market purchases is unlikely to achieve this interest. Contracts, on the other hand, can specify planting dates (and thus delivery dates) as well as total quantities to be delivered.

Contracting reduces much of the uncertainty that would exist if the company simply bought crops on the open market, and gives the company some control over the production process (for example, over the variety grown). There is no reason, of course, for a company not to use more than one method of obtaining its supplies, and some companies use a combination of company farms, contract growers, and open market purchases.

Many contract-farming schemes in less developed countries are multipartite arrangements involving private firms (usually foreign, but occasionally local), the host-country government, and international aid or lending agencies, such as the U.S. Agency for International Development (USAID), the World Bank, or the Commonwealth Development Corporation (CDC). The CDC has been particularly active in this type of scheme. In one common variation, a national development bank provides growers with credit for the purchase of fertiliser, seeds, and other inputs. At harvest time, the company pays growers the contract price, but takes off a sum that goes to the bank to repay its loan to the grower. In this system, private companies avoid the problems of assessing creditworthiness and prosecuting defaulters. In some cases, government agencies provide inputs or technical assistance.

Contract-farming schemes tend to be of two types. One type produces traditional tropical commodities, such as sugar, rubber, or oil palm, which tend to be produced at lowest cost on large tracts of land. Contract-farming schemes in such commodities usually involve a large number of growers, tight central control, and provision of numerous services by the central processing unit (for example, irrigation, harvesting, and aerial spraying). There is usually heavy involvement of external donors in these schemes. Many such schemes originated as resettlement schemes. In extreme cases, some schemes are, in effect, disguised plantations, for example, the

Papua New Guinea cardamom scheme. These large projects, often referred to as outgrower schemes, are particularly common in Indonesia and Malaysia (rubber and oil palm) and in Africa (oil palm, sugar, and tea).

Another type of contract-farming scheme, usually on a smaller scale and with more private-sector involvement and less tight centralised control, is frequently used in fruit and vegetable production, particularly in Central America and Thailand. Most frequently, this type of scheme entails the export of high-value items, such as asparagus, cucumbers, melons, or strawberries, with the company providing quality control, brand names, and marketing channels. Business-oriented growers, cooperatives, and individual small fanners are all involved. Fresh vegetable exports from Guatemala are an example of such schemes.

Total developing-country employment in contract farming of these nontraditional crops is much less than in traditional crops, such as bananas and sugar. However, there is some evidence that contract farming of nontraditional crops is expanding at a faster rate and that these labour-intensive products are more promising outlets for small farmers. Several aspects of contract fanning impinge on food production and consumption by growers, their families and employees, and other segments of the population.

Income Generation

One of the key features of contract farming is that it provides farmers with access to markets that would not otherwise have been available to them. Without the quality control and tight coordination offered by contract farming, it is frequently unlikely that smallholders would be able to sell perishable goods overseas through open market sales. The most significant income increases have been generated in those schemes in which smallholders gain access to lucrative export markets for labour-intensive luxury crops. Important income increases have resulted from traditional crops as well, however, particularly tea.

Furthermore, contract farming in cash crops often brings significant changes with respect to both size and frequency of payments to the recipients. It is difficult to generalise about the direction in which contract farming will shift the payment system-this is highly case specific. A contract-farming crop with a fairly continuous harvest and payments may replace one harvested and sold once a year in the local market, or the reverse may be true. It is also possible for contract-farming schemes to separate the payment system from the harvest cycle by providing weekly or monthly advances that resemble a wage, as often occurs in outgrower schemes. This is a possibility in contract-farming schemes; in open market sales it is not, at least not without recourse to the capital market.

Dislocation of Alternative Crops

Contract farming could reduce food production if a new contract cash crop displaced a previously grown food crop. Evidence suggests that such displacement does not frequently occur when farmers are allowed to make their own decisions. In such cases, the contract crop tends to displace a less profitable cash crop rather than a food crop. In centrally managed schemes, these decisions are often made by the scheme authority. Many schemes require growers to maintain a certain acreage of food crops, though a few have gone in the other direction and required growers to specialise in the contract crop.

A series of African case studies found that when land was fairly abundant and farmers had the freedom to diversify their crops and income sources, food supply was adequate and no special measures were needed from the scheme authorities. In Swaziland, for example, households produced much of their food on family plots outside the scheme, allowing them to use their scheme plots for sugar. Reduction of food supply was most likely to be a problem in areas where availability of land was a constraint.

Schemes that did restrict production of noncontract crops tended to encounter serious problems, not only in food production, but also in maintaining adequate levels of recruitment and participation in the schemes. For example, some schemes in Zimbabwe restricted farmers by allowing them to grow only the contract crop on their plot or by admitting only full-time farmers, thus forcing them to forgo off-farm income. In each case, farmers either evaded the restrictions (sugar and cotton schemes) or the scheme became unviable (tea schemes).

Local agroclimatic and market conditions are sometimes more suitable for cash or export crops than food crops. In such cases, there can be significant welfare gains from trade. For example, a cooperative of former banana workers was formed in Honduras in the early 1960s to farm land formerly given over to bananas. The cooperative experimented with many food crops but found that they did not grow well in their soil and drainage conditions. On reverting to bananas (the export crop), the cooperative's income increased to levels far above those of food crop producers.

A key feature of contract farming, which bears on production response, is risk sharing and risk reduction. In fact, contracting is fundamentally a way of allocating risk between the company and its growers. The latter assume most of the risks associated with production, while the former assumes the risks of marketing the final product. Total risk is reduced relative to a noncontract situation of that crop. However, it is unclear whether total risk for the farmer is reduced, since nontraditional crops are inherently risky; when the addition of the contract crop to the farm's product increases total risk for the farm, it may induce a farmer to increase the percentage of subsistence requirements.

Multiplier Effects for Employment, Infrastructure, and Market Development

Local governments often favour contract farming in the belief that it will produce greater spillover or linkage effects with the local economy than would plantation production. Studies have found significant variations in this respect. For labour-intensive fruit and vegetable crops, a multiplier is clearly present in the great expansion of daily farm labour employment made necessary by the new contracted crops. This was found to be the case with cauliflower and broccoli cultivation in Guatemala.

There are also significant new employment opportunities in the transport and processing sectors. Again, the effects of this employment and income generation on food supply will depend on income elasticities of demand by labourers and on domestic supply response to that demand. For traditional export crops, and in cases where a highly mechanised and centralised production system is transferred to large outgrowers, the situation is different. Here, the nominal transfer of legal responsibilities via a contract does little to change the economic imperatives of the production system.

In the Central American banana industry, for example, the linkage effects of local production are not much greater than those provided by transnational corporation production; nearly all of the inputs used by associate producers are provided by exporting companies. Moreover, as Ellis shows, the linkage effects of any form of banana production are very slight. In addition to direct employment effects, some large contract-farming schemes in remote areas have had broader rural development impacts. To some extent, these schemes have acted as growth poles. This has occurred in sugar schemes in western Kenya, tea schemes in Tanzania, and in a frontier asparagus scheme in Peru: all schemes have performed well in terms of opening up underdeveloped areas of the countries in which they are located.

Construction of roads and other infrastructure and expansion of interregional trade have been some of the direct consequences of the establishment of contract-farming schemes. There is some evidence that these rural development effects could have been greater if more deliberate planning had been carried out and local resources had been used more frequently. The chief relevance of contract farming as an organisational form in this respect is that the market, on its own, was not opening up these regions.

Market development also results from the production process itself in some environments: the reject rate in fruit and vegetable export operations, which usually rely on contract farming, often reaches 50 percent. These rejects-often set aside merely for reasons of ripeness or size-can be sold in local markets for a fraction of the price paid by industrial-country consumers. The produce has nutritional value and can complement the traditional starchy diet of low-income consumers.

In some cases, exotic vegetables are not well accepted initially but catch on as their characteristics and preparation methods become better known. In Guatemala, for example, reject cauliflower and broccoli are so widely and cheaply available that they have become a nutritious staple of the poorest people; the leaves and stalks are also used as animal feed and as an organic fertiliser for the noncontracted (mainly subsistence) crop fields.

Effective Extension Services

The contractual relationship between growers and a processing company in contract farming provides the latter with the assurance that it can appropriate a share of the benefits from the investments it makes in production at the farm level. This is most apparent in the areas of extension and input provision. The company has a direct interest in providing effective extension services because it wants high-quality, lowcost produce. Public extension services have no such incentive and regulate their performance in accordance with bureaucratically defined criteria (for example, number of farmers served, quantity of inputs distributed). These criteria are much less effective in assessing performance and providing incentives than the profit-related criteria used by contracting companies.

A pirori, then, one would expect the quality of extension provided in contract farming to be superior to that found in purely public or market-oriented systems. In most private contract-farming schemes, extension tends to deal only with the contract crop, although some of the multipartite schemes in Africa use a multicrop approach. Some of the production techniques learned in contract-farming schemes are highly crop specific and are not transferable to other commodities.

Management skills learned through participation in an agribusiness scheme are more widely applicable, however, and include accounting practices, negotiating skills, and awareness of the importance of quality, characteristics of export markets, and contract provisions. Generally speaking, there tends to be some transfer of contract-farming-induced production and management skills to other cash crops and to the farm enterprise as a whole.

The situation is similar with respect to input provision. The volume and timeliness of delivery of agrochemicals should be close to optimal in a contract-farming situation, since the company has a secure means of ensuring repayment of in-kind credit through deductions from crop payments. Chemicals applied can have residual effects on adjacent or rotated food crops. In addition, inputs delivered for use on the contract crop are occasionally (and illicitly) diverted to food crops, depending on the farmer's estimate of relative marginal returns. Irrigation water provided through contract-farming schemes has also been used for secondary crops, as in some African sugar schemes that support the production of vegetables for local markets.

On most schemes, basic agronomic research is fairly limited; the emphasis is on disseminating existing technologies rather than on developing new ones. There is, however, unexplored potential for contractfarming schemes as sites for the testing and introduction of new food crop technologies. At present, much basic and adaptive research is carried out on experiment stations and tested in study villages under fairly tight control. From there to nationwide dissemination is a big leap; contract-farming schemes could provide an intermediate step. Other aspects of the contracting relationship may have moderate to weak effects in either a positive or negative direction.

The priorities, then, are first, to maximise the income-generating effects of contracting schemes; second, to moderate those secondary aspects that have potential negative effects; and third, to increase the secondary positive effects. Measures to improve the financial viability of schemes are of particular importance. These measures include setting appropriate pricing policies, rewarding risk taking by private companies in new crops or regions, and improving the autonomy and accountability of parastatals. Since much income is generated as a result of access to export markets, identification of potentials of nontraditional exports, studies of marketing channels, and market promotion efforts will be useful.

Payment systems can be modified to provide smaller, more frequent payments. Where feasible, project authorities should sign contracts and disburse payments to household members who actually carry out the work. Where these are women, there will likely be a positive effect on food expenditures. Farmers should be given as much freedom as feasible in managing their enterprises, particularly with respect to choice of crop mix and off-farm activities. Restrictions on noncontract crop activities should be avoided. The multiplier effects of contract farming can be maximised by encouraging project authorities to plan for development of investment opportunities into which growers can channel their new income. Greater use of local resources in transport, maintenance, and manufactured goods provision can also make a contribution.

Extension services should be designed to provide learning effects that go beyond production of the contract crop. It is unlikely that the debate over single-crop versus multicrop extension systems will ever be resolved: it is difficult to assess the trade-off between the technical superiority that comes from specialisation and the efficiency in delivery that comes from multicrop extension. Contract farming studies, however, tend to give more support to the latter. Farmers seem to prefer the farm management approach provided by multicrop extensionists, and it could be argued that specialised extension services are often not feasible in very poor countries.

These skills are more readily transferable to food crops than are production techniques. Management skills are most likely to be developed in schemes where producer prices closely reflect quality and final market prices; where farmers receive detailed accounts of the company's payments for crops and deductions for inputs; and where farmers are given substantial responsibility for managing their operations, rather than operating within a scheme where control is highly centralised.

11

Consequences of Contract Farming

Specialty grains, feeder livestock, even fruits and vegetables, are being produced under contract, and have been for some time. Basically, the concern is a tilt in market power with a possible shift in bargaining power as input suppliers and output processors (and first purchasers otherwise) gain greater economic power, undoubtedly at the expense of producers. It is assumed that the governing policy goals for the food and agriculture sector will continue to include: (a) availability of an abundant supply of food, at reasonable prices; (b) maintenance or enhancement of the productivity and environmental integrity of natural resources; and (c) a prosperous and productive economic climate for producers (including family farmers).

Mergers, alliances, and various other forms of arrangements are reducing the number of players in input supply, particularly in seeds, and increasing the level of concentration. While the level of mergers, alliances, and consolidations is not a completely reliable indicator of competition, the fact that nearly $15 billion of such amalgamations has occurred over the past three years, some at price levels difficult to justify under present economic conditions, suggests that (a) some are discounting revenue from a pot at the end of some unknown rainbow, (b) irrational behavior is being displayed, or (c) some acquiring firms are assuming that a greater share of the world's food bill can be claimed by seed suppliers.

But increasing levels of concentration among firms do not tell the entire story. The revolution in ownership of germ plasm, the feature of cells that determines the characteristics of offspring, also is moving rapidly toward concentration in a few hands. Increased concentration is also leading to control by a few firms over the major processes by which genetic manipulation occurs, thus enabling those controlling the technologies to block use by other firms.

This development is partly related to the changing role of the land grant universities, partly to the ability in recent years to manipulate germ plasm through genetic engineering, and partly to the consequences of the ability to obtain a monopoly-like position over unique life forms and over the process of genetic manipulation.

For decades, the land grant universities developed the basic genetic lines and made those lines available to the seed industry. Because of limitations on university funding and the near-revolution in genetic engineering, the private sector several years ago began pouring more money into basic research. Developments have progressed to the point that the payoff from research and development funding can no longer be used to compare the present with prior periods. Payoffs are expected to flow more readily than when biotechnology was in its infancy.

The advent of genetic engineering meant that scientists could manipulate genetic composition-not through conventional crop breeding techniques, but through laboratory procedures-to change the genetic makeup of plant and animal life. That has produced herbicide-resistant crops, for example.

Effect of Contracts

An important question is the effect concentration in the seed business and control by the few resulting firms over germ plasm will likely have on contract negotiations with producers. It depends on the options open to producers who don't like the terms of contracts offered to them. With numerous contract possibilities available from input suppliers, each offering inputs of roughly equal productivity and cost, the answer is perhaps "not much."

But if there are just a few options, with the next best offering a much less attractive set of inputs in terms of cost and productivity, such as when a variety of seed is developed with significant yield premium over otherwise competitive varieties, the answer is "take what you're offered." A greater proportion of the value of the yield premium is expected to be captured by the seed supplier under those conditions than has historically been the case. The outcome is likely to be a tilting in the terms of contracts in favor of the input supplier. The division of revenue from production would be expected to shift over time in favor of the party with the monopoly or near-monopoly position.

Seed companies and other input suppliers can be expected to drive the best possible bargain-which means, in the case of seed, capturing the greatest possible percentage of the value from any yield premium. The outcome would be a smaller share of the revenue from production going to the producer, resulting in less compensation to the producer and less to capitalise into land values. Seed companies would end up with a larger share of the pie with more to capitalise into the stock of the input supply firms. Even if unique corn derivatives produce revenue of $2 million per acre, it's fairly clear that whomever holds the rights to the technology involved will capture the lion's share of the revenue, not the producer.

A good argument can be made that this perception of potential profits in the future is part of what is driving the intense push toward concentration in control over germ plasm now occurring. Thus, a major issue is whether a shift in market power occurs between input suppliers and producers, whether that shift in market power is translated into enhanced bargaining power, and whether the enhanced bargaining power is employed to siphon a greater proportion of the economic return generated by the sector into the hands of input suppliers.

The negotiating power of seed firms could well have other impacts.

— In an effort to control the germ plasm more completely, seed companies are likely to negotiate for ownership of the product with the producer under contract having only a contract right to payment, short of ownership of the crop or livestock involved.

— Similarly, the contract may contain what would appear at first glance to be an attractive feature-the input supplier bearing the price risks.

These seemingly innocent shifts would mean, however, that the economic position of the producer would be transformed from that of a risk-taking entrepreneur into a relatively riskless world of fixed compensation. Thus, a shift not only of compensation would occur in favor of the input supplier, but also a shift of management functions in the same direction. The outcome would be reminiscent of the limited role played by growers under broiler contracts.

Barriers to Entry

In general, one would expect high-handed economic behavior by near-monopolists to be met by entry of new competitors attracted by the generous terms of contracts in favor of the input suppliers. And that would likely occur if entry were possible. However, barriers to entry may be fairly high.

- — One barrier is capital needed to mount the kind of research effort needed to maintain a product flow similar to that of the firms pressing for monopoly-like concentration levels. The capital needed is very substantial.
- — Also, existing patent and plant variety protection may mean that potential competitors are frozen out of competition as a practical matter for the duration of the patent or PVP certificate, or the duration of a patent over processes by which genetic manipulation occurs.

Small Seed Firms

A major issue is whether smaller seed firms are likely to be able to acquire germ plasm and thus remain viable. Certainly the small firms have remained surprisingly healthy in recent decades as performance traits of the varieties and hybrids developed by the larger firms have tended to outdistance the performance of seed marketed by small firms.

But the era of transgenic hybrids produces both the incentive to maintain greater control over high-performing germ plasm and the technology and resources to challenge those who manage to obtain the germ plasm in clandestine ways. The larger firms may acquire some smaller firms to complete their distribution networks, and licensing germ plasm for a fee may well occur. However, it is unlikely that the dominant firms will generate additional competition by licensing to smaller firms.

Indeed, with the smaller firms predictably unable to maintain access to higher performing germ plasm, the price of lower performing seed varieties and hybrids is expected to reflect the economic disadvantage inherent in the lower performing varieties. At some point, many if not most of the smaller seed firms that are unaligned with the dominant firms will be unable to survive economically.

Collective Action by Farmers

One possible strategy for farmers is to forge alliances among producers. The push to achieve such countervailing power was the driving force behind the formation of labor unions a century ago. Historically, however, farmers have been unwilling to accept such a disciplined approach to achieving bargaining power. Producers and consumers are bound together by economic laws which they did not make and which they cannot repeal. Between these two are powerful agencies whose only interest it is to take such toll as they may, as products are passing from producer to consumer. These agencies, by reason of their financial connections, exercise an influence far greater than is warranted by their numbers or the service they perform.

Role of Institutions

Arguably, what is likely to emerge over the next few years is a heightened awareness of the efficacy of institutions in limiting or constraining economic activity. To the extent that institutional intervention is successful, a major concern is how to keep institutions in adjustment with changing economic circumstances. Markets reflect changes day by day, minute by minute. Yet, institutions tend to remain in place, frequently producing economic rents for some, until sufficient momentum is generated to effect change. To a considerable degree, institutions limit (as well as facilitate) market operations, but without the same self-adjusting features as markets.

Vertically Integrate the Sector

The moves made by the major players, particularly the seed companies, could lead one to conclude that the objective is to vertically integrate the sector. Such an objective could be pursued for several reasons: (a) to gain and maintain greater control over patented products or products subject to intellectual property protection otherwise, (b) to apply economic pressure on producers to relinquish functions in favor of the integrator (such as risk management) or to merely provide an opportunity for risk to be off-loaded onto the integrator, (c) to enhance profits of the integrating firm, or (d) to achieve greater market share on an assured basis. Some readily concede that an objective of vertical integration is one of the forces driving the push toward a contract-based agriculture.

12

Contract Farming in Agri-food System Development

Since contract farming is by no means a novel issue, there should then be a specific reason for such heightened awareness and this is most plausibly associated to the remarkable pattern of changes and trends affecting agri-food systems worldwide. Spurred by drivers such as technological developments, demographic changes, changing consumer preferences, trade liberalisation and financial capital mobility, food and agricultural systems are being forced to adapt and modernise. Increasingly, such systems are becoming organised into tightly aligned chains and networks, where the coordination of production, processing and distribution activities is closely managed.

In these modernised systems, the once dominant role of spot markets as a mechanism to harmonise transactions is being replaced by alternative forms of vertical coordination, including strategic alliances, full ownership and contracts. As a consequence, supply chain management principles, which already impacted the organisation and performance of manufacturing and retailing internationally, have found in the agri-food sector a new, fertile ground to be put into practice. These transformations, and the organisational responses thereof, are creating challenges and opportunities for producers, processors, wholesalers, retailers and other supply chain actors. Small farmers in developing countries, in particular, are perceived to be especially vulnerable to the changes.

Modern organisational arrangements in agri-food systems might promote the emergence of power imbalances and unfavourable terms of trade in the transactions between smaller-scale chain actors and the larger players which typically exercise the leading coordination role in a managed supply chain. But these perceptions notwithstanding, contract farming is being promoted by governments and development agencies as a coordination mode that can facilitate the integration of small farmers into supply chains. This also accounts for the rising interest, earlier pointed out, in the conceptual foundations and in the practical implications of contracting in food and agriculture.

Changing Agri-food Systems

Agri-food systems around the world are undergoing a distinctive transformation process. Changes are being observed in all segments of the production-distribution chains, as well as in the institutional environments in which they operate. We are witnessing a phenomenon that has been called by some as the "industrialisation" of world agriculture, a metaphor that highlights the departure from the traditional ways in which food and fibre were produced and marketed to a mode under which farms tend to increase in size (and reduce in numbers), adopt production processes akin to the practices of manufacturing enterprises and develop closer ties to processors, retailers and other partners in the supply chain.

At the institutional level, we are observing, inter alia, the enactment of stricter quality and safety regulations, the emergence of private grades and standards and the reduced role of the public sector in traditional supporting activities and services provision. These general trends, while more pervasive in the developed world, are also present in developing countries. Moreover, their effects are not bound by geography: in a globalised world, where supply chains and networks are intertwined across countries and continents, the consequences of the changes are far-reaching. The general perception is that the reshaping of agrifood chains is being driven by a number of demographic, technological and economic related factors.

Growth of Population and Income Growth Urbanisation

Population and income are the main factors affecting the demand for foods. The global population reached 6.4 billion in 2004 and grows by about 80 million every year, with fastest rates in Sub-Saharan Africa and West Asia and slower growth, even population decline, in most of the industrialised world. These trends are expected to continue, particularly in the developing world, where the projected population growth rate is almost 2% per year until 2010.

As population grows, the share of inhabitants living in rural areas tends to decline sharply (Figure 1). It can be inferred from Figure 1 that by 2015 more than half of the world's population is expected to be urban. Moreover, the UN estimates that in the next 25 years almost all of the world's population growth will be concentrated in urban

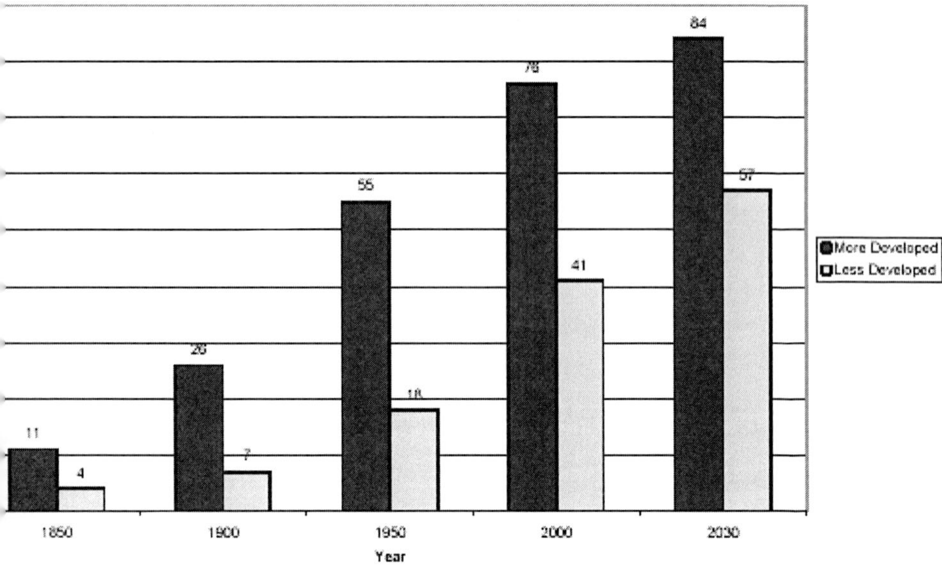

Figure 1. Urbanization rates 1850-2030

areas of developing countries. This shift of population from rural to urban areas is fuelled by economic development and as such stimulates income growth. In fact, the latest projections by the World Bank show income per person increasing in all regions of the world (Table 1), with particularly high rates in the Asia-Pacific area.

Table 1. Projected Growth in GDP per Capita (% per year)

Region	2005	2006	2006-2015
East Asia and Pacific	6.2	5.7	5.3
Europe and Central Asia	5.6	5	3.5
Latin America and Caribbean	2.3	2.3	2.4
Middle East and North Africa	2.8	2.5	2.6
South Asia	4.7	4.4	4.1
Sub-Saharan Africa	1.6	1.7	1.6

With few regional exceptions, this is a trend that has been recurrent in the recent past: incomes have grown and will continue growing in the foreseeable future. In average, developing countries should experience a growth rate of 4.4% per year in GDP per capita between 2006 and 2015. The combined effect of such income and demographic patterns in agri-food systems is already being manifested, mostly by quantitative and qualitative changes in the demand for agriculture and food products. FAO forecasts that the overall demand for foods will be approaching 3000 kcal/person/day in 2015 and will exceed 3000 by 2030, rising from 2803 kcal/person/day in 1997/1999. The diets are also changing, with increases in the proportion of calorie intakes being derived from livestock, oil crops and sugar. Consumption of high value animal protein foods, for instance, has been rising steadily and is expected to increase by 5% yearly until 2007 and by 8% per year in the subsequent 5 years.

In sum, there is a need to produce more food and to diversify the composition of the traditional food basket. From the qualitative standpoint, consumers are also becoming more demanding and more concerned with non tangible attributes of foodstuffs. Consumers keep basing their choices on traditional food purchase drivers such as price, freshness and wholesomeness, but increasingly they demand information about and have their buying patterns influenced by aspects related to health and safety, environment conservation, animal welfare and social responsibility, to name a few. A shift from commoditisation to product differentiation is thus being observed.

Finally, the impacts of the demographic and income trends are additionally being felt in the ways foods are being processed and distributed. More urban and affluent consumers are demanding foods that can be conveniently prepared, such as frozen, pre-cut, pre-cooked and ready-to-eat items. They are also spending greater shares of their budgets in the "food-away-from home" category. Processing and distribution systems have been adapting, as suggested by the growing importance of agroprocessing and modern food retailing, especially in the developing world, in terms of contribution to growth of domestic product and employment generation.

Trade Liberalisation and the Mobility of Capital Flows

The globalisation process is characterised by the free flow of goods, services, capital and knowledge among nations. Among its consequences are the increasing internationalisation of firms and the integration of global markets, via trade mechanisms. Trade liberalisation and the mobility of capital flows are at the root of these processes and have affected the organisation and performance of agri-food systems worldwide.

Trade of agricultural products has not only been growing in absolute values, but with liberalisation the composition of world agri-food exports has altered. By the end of the 1990s, the share of processed agricultural products trade in relation to commodities had increased from 24.7 % in 1970 to 58.2%, a trend that had similar patterns in developed and developing countries. Traditional export commodities, such as cocoa, coffee and sugar had their relative importance reduced, whereas increased trade occurred in fruits, vegetables and dairy products, a large portion of which in processed form.

More trade and enlarged markets for higher valued products bring about intensified competition and the competitive pressures stemming from increased exposure to international markets affect food chains in diverse ways. A major one is the need for improved technical efficiency in producing, processing and distribution, leading to reduced product costs and hence to potentially larger market shares.

A further impact is the need for adaptations in the so-called enabling environment, which includes the set of norms and regulations that affect the way agri-food systems operates (e.g. health and safety regulations related to foods, norms governing industry entry and exit, labour laws, foreign exchange legislation, property rights, trade licensing, etc.) plus the set of support services (information, financing, insurance, research and development, etc) that facilitate transactions to take place.

A third influence has to do with the organisational changes necessary to ensure that chain transactions are efficiently coordinated across firms locally and across borders, often calling for the adoption of supply chain management principles.

Improvement in Transportation and Logistics

A study published by the Worldwatch Institute called the attention to the fact that the distances between food suppliers and the consumer have been growing. In the US, food typically travels between 1,500 and 2,500 miles from farm to plate, as much as 25 percent farther than in 1980. The average distance travelled by the components of a typical meal in the UK is believed to be above 3000 km. These figures, while partially reflecting the growth of international food trade earlier discussed, also suggest that advances in transportation technologies are a further factor influencing the reshaping of food systems.

Advances in containerisation, reefers, controlled atmosphere, intermodalism, vessel sizes and speed, fuel efficiency and satellite navigation systems, among others, are contributing to reductions in freight costs. Air shipment has become a standard international transportation mode for perishable products—loads of mangoes from Brazil, flowers from Kenya or fresh vegetables from Ghana, for instance, can be cost effectively sent overnight to Europe. In sum, transportation and other logistics advancements enabled long-distance sourcing and allowed the diversification of the supply base of food retailers and processors. It also favoured the growth of farming in sectors such as livestock rearing or others that depend on feeds or raw materials not locally available.

In domestic markets, improved transportation technologies are also favouring changes in food systems. An example is the dairy industry of Brazil, where after refrigerated bulk milk collection was introduced in the 1990's, the milk collection radius expanded from an average of 30 to 50 km to more than 300 km. The new logistic system promoted the intensification of competition among dairy firms and also displaced small farmers that could not make the investments required to adapt. Other examples can be drawn from retailing, where consumer mobility enabled by improved transportation has influenced the pattern of store location and sizes.

Improvement in Information and Communication Technologies

The path of developments in information and communication technologies has been dramatic, affecting all segments of agri-food systems in both direct and indirect ways. In 1965, when the famous Moore's law was proposed, a single transistor cost more than one dollar. The first microprocessor had 2200 transistors. Today, a state-of-theart computer chip has well above 1 billion transistors that cost less than 1/10000of a cent per unit. Computers became much faster, smaller and less expensive and with this thrust information and communication technologies (ICTs) became widely adopted in all areas of economic activities.

For agri-food chains, ICTs allow fast and cost effective collection, storage and retrieval of data at its different stages. Information can then be readily exchanged among organisations and fed into managerial systems that permit better planning, control and decision making. Supply chain management in the agri-food industries is today heavily based in such information exchange processes. Consumer purchases, registered at the check-out counters of retail stores, immediately trigger inventory adjustments and replenishment orders from upstream suppliers. Supply chain efficiencies yielded by these systems are believed to be the major driver of the dominance of large retailers, such as Wal-Mart, in the food distribution business.

Computer and communication technologies also allow better consumer responsiveness and compliance with food safety and quality regulations, as illustrated by the developments in bar codes and scanning that enabled product tracking and tracing. Veal consumers in Europe, for instance, can use the Internet to enter the bar code appearing in product packages and have immediate access to information on the production and processing steps followed from the farm to the supermarket. With bar codes and scanning, events of food contamination can be rapidly communicated down and upstream a chain, so as to isolate the source of the problem and ensure that consumers are not unduly exposed to health hazards.

Other significant developments are: the precision farming technologies that are contributing to gained efficiencies in farm productivity and input use; the automated devices used to improve controls in agricultural and post-harvest operations; the spread of mobile phones, which are contributing to increase market transparency and reduce transaction costs; the emergence of radio frequency identification systems (RFID), with the potential to further improve logistical management and the use of e-commerce in food and agricultural markets.

Improvement in Biotechnology

Technological progresses in production and processing have historically allowed the growth of global food supplies to outpace that of population. Developments in fertilisation methods, crop improvement and in animal genetics and nutrition have contributed to the increase of agri-food output per unit of land area, whereas improved post-harvest techniques permitted the conservation of products and their marketing under lower geographical and timing constraints. More recently, biotechnology is seen as the new step forward in the quest for food security. Notwithstanding the perceptions of potential risks, its development promises substantial benefits to all stakeholders of agri-food supply chains.

Biotechnology refers to the application of biological research techniques to the development of products that improve agriculture and human health. In agriculture, it involves the application of markers to assist in plant breeding, the techniques of tissue culture to allow the multiplication of plant materials and the genetic engineering, or manipulation of an organism's genome by introducing or eliminating specific genes.

With regard to agri-food systems organisation and performance, biotechnology has the potential to negatively impact market structure. As illustrated by the recent experience in the US, in key segments such as crop seeds and chemical inputs, the high costs of research and development, the high regulatory costs and the time bound monopoly conferred to holders of property rights are leading to industry consolidation, via mergers and acquisitions. Market power allows practices such as the requirement that farmers sign contracts impeding the reuse of seeds and limiting the type of herbicides they can use. Another impact is related to the rapid spread of genetically modified (GM) crops in some areas of the world. According to the International Service for the Acquisition of Agro-Biotech Applications (ISAAA), the estimated area with GM crops in 2004 was 81 million hectares, up from 67.7 million hectares in 2003. These crops—mostly soybeans, maize, canola and cotton—were grown by 8.25 million farmers in 17 countries, most of which in the developing world.

GM based products, although widely accepted in the USA, are still a controversial issue for consumers in most areas of the world. Legislation regulating marketing of foods increasingly requires that special labelling is employed to inform consumers as to the presence of GM ingredients. Agri-food systems have to respond, by putting in place appropriate procedures and establishing chain wide standards to facilitate tracking and tracing. Clearly, the enactment of stringent traceability regulations is a factor that favours the development of tightly aligned supply chains.

In sum, as agri-food systems are reshaped in response to the trends here discussed, becoming more competitive and globalised, the ways in which supply chains are coordinated have to be adapted accordingly. Better synchronisation of the vertical stages of agri-food value chains becomes necessary, as a way to lower costs by improving productivity, improve and ensure quality throughout the chain, control risks associated with markets and food safety and enhance responsiveness to demand.

Theoretical Approach for Contract Farming

Some authors trace the principles of contract farming back to the 19century, when the mechanism was used in the United States for processing crops such as sugar beets and peaches, and in Taiwan, for sugar production under the Japanese colonial rule. Its use later expanded into many food and fibre sectors, particularly in the US and in some Latin American countries supplying the US markets. Recently, contract farming became even more widespread. The United States Department of Agriculture revealed in a recent report that contracts now govern 36% of the value of US agriculture production, up from 12% in 1969 and are the primary means of vertical coordination in sectors such as hogs, broilers, pork, sugar beets, fruits and processing tomatoes.

In developing countries, contracts are also being increasingly used. In Brazil, 75% of poultry production is coordinated via contracts, whereas in Vietnam there are indications that 90% of cotton and fresh milk, 50% of tea and 40% of rice are being purchased by enterprises though contracts.

Contract farming can take different formats and some typologies for it have been proposed. Minot cites R. Mighel and L. Jones as originators, in 1963, of the widely used classification of contracts into three kinds, namely "market specification", "resource providing" and "production management". In the first modality, the transaction between growers and buyers is agreed on terms of what to be produced (product and quality attributes) and what are the commitments for future sale (timing, location and price). The second modality adds the provision of farming inputs to the former contract type.

Beyond specifying what to produce and what the conditions for marketing are, in-kind credit is offered via the provision of key inputs, often with cost recovery upon farm product delivery. Finally, under production management contracts growers agree to follow precise technological guidance on how to produce. But regardless of the typology, the general term "contract farming" refers to a particular form of supply chain governance adopted by firms to secure access to agricultural products, raw materials and supplies meeting desired quality, quantity, location and timing specifications

Spot markets, or open markets, are the simplest form of vertical coordination. Transactions are coordinated by prices only and there is no continuing obligations among those engaging in them. For markets where the conditions resemble perfect competition (i.e. many buyers and sellers, homogenous products, perfect information, etc.) spot markets can be efficient coordination mechanisms. The chains for agricultural commodities are typically governed by spot markets. Situated in the opposite extreme of the vertical coordination choice continuum, full vertical integration refers to cases where a firm exercises centralised ownership control over two or more different stages in a chain. Ownership, of course, ensures complete power over decisions regarding product attributes, location and timing of delivery. The supply of agricultural ingredients to the pharmaceutical industry is often cited as an example of transactions where vertical integration tends to be the governance norm.

Contracting is an intermediate mode of coordination, whereby the conditions of exchange are specifically set among transaction partners by some form of legally enforceable, binding agreement. The specifications can be more or less detailed, covering provisions regarding production technology, price discovery, risk sharing and other product and transaction attributes.

The contributions of transaction cost economics, is perceived as particularly significant to the case of contract farming. Simply stated, transaction costs are the costs incurred when a firm engages in an exchange process. They include the costs occurring before a transaction takes place, such as obtaining information and negotiating the exchange conditions, and the expost costs of monitoring and enforcing the transaction terms. When choosing a governance mode, firms seek to minimise transaction costs and in this process they have to examine the characteristics of transactions related to asset specificity, uncertainty and frequency.

Asset specificity refers to the degree in which an asset can be economically transferred to alternative uses. Land, for instance is a non specific physical asset, whereas a milk parlour is a highly specific one. Site specificity, on the other hand, exists when buyers and sellers base their location decisions on the need to reduce transportation costs in their exchanges—the location cost advantage is lost if the commercial bond that motivated it is broken. Time specificity is associated to the timing of delivery and its effect on product value—more perishable products, for instance, lose value if not timely marketed.

Human capital specificity arises as a consequence of accumulated knowledge by participants in production processes. The specific skills acquired might have lower value in alternative uses. The higher the degree of asset specificity, the higher will be the exposure of the transaction partner who holds the asset to exploitative or opportunistic behaviour from his counterparts. Hence, a high degree of asset specificity drives transactions away from spot markets, towards tighter alignments in the supply chains.

Uncertainty in transactions arises from three main sources. One is the behaviour of those engaged in the exchange, which can be influenced by opportunism. Reneging on previously agreed terms, nondisclosure of undesirable attributions or distortion of information are some characteristics of opportunistic behaviour that generate uncertainty on a transaction. The second has to do with uncontrollable factors, such as technological changes, acts of nature or consumer preferences. Finally, a third source is related to the inability to control decisions and plans made by others, which may affect attributes such as product quality and timing of delivery.

As uncertainty increases, firms have more incentive to seek control over the transactions, thus moving from spot markets to more vertically coordinated governance modes. The frequency of transactions is directly correlated with the incentive to opt for spot markets. When transactions are frequently performed, buyers and sellers tend to engage in longer term—but not necessarily exclusive—business linkages, thus reducing the scope for opportunistic behaviour. Information about buyers and sellers tends to be disseminated by the repeat transactions, reinforcing the motivation for the continuance of the open market transactions. Note that frequency of transactions is a different concept from frequency of production. The latter refers to an internal operation, being related to production costs, rather than transaction costs.

Obviously, when a firm produces a certain type of product only sporadically, it will have little incentive to engage in close coordination modes with partners in the supply chain. Table 2 summarises the theoretical basis for the selection of a mode of governance, in accordance with transactions cost theory. Negative and positive signs are used to denote the relationship between the characteristic and the likelihood of adoption in the respective governance mode.

Table 2. Choice of the governance – A summary

Characteristic	Open Markets	Tightly Aligned Chains
Asset specificity	-	+
Uncertainty	-	+
Frequency of transactions	+	-

Although this theoretical base is useful to discuss the observed expansion in the utilisation of contract farming and the movement towards the development of more tightly aligned supply chains in light of the changes taking place in agri-food systems, it is at best an incomplete framework in terms of its explanatory and predictive power in empirical analyses. The contractual instrument, in principle desirable to reduce the degree of uncertainty caused by opportunistic behaviour, was deemed ineffective because of an institutional failure.

Market and institutional failures, such as the one identified by Bando, can affect the transaction characteristics summarised in Table 2 in positive and negative ways. Hence, authors such as Minot, Key and Rusten and Kirsten and Sartorious, have focused their discussion of the theoretical basis for contract farming on the examination of such failures. They discussed asymmetries in production and marketing information, as well as the imperfection in markets for credit, inputs and agricultural support services and indicated that these failures account for increases in transaction costs and thus provide incentives to increased coordination in the transactions, leading to contracting or to full vertical integration.

Minot goes further to indicate that scale complementarities would influence the choice between contracts and full integration—if economies of scale, location and or scope are complementary in adjoining links in a supply chain, then a motive exists for integration. His example is the integration observed in the US among feed and livestock operations. The bulkiness of feeds favours location of processing plants closer to livestock operations, which in turn are more efficiently operated on large scale. Hence, vertical integration is favoured by both location and scale complementarities.

In sum, the study of vertical coordination in agri-food systems provides the backdrop for our understanding of supply chain organisation and performance. Within this general setting, and in spite of known shortcomings, transaction cost theory is the conceptual framework of choice in recent analyses of contracting farming issues. We turn now to examine the potential advantages and disadvantages of this mode of governance for the actors involved in the transactions.

Advantages and Disadvantages of Contracting for Agri-Food Chain Actors

A useful instrument for public policy design is offered by stakeholder analysis, which among other things tries to identify who stands to win or lose from a policy choice and assess what is the extent of their potential gains and losses. Private decisions on strategic choices, on the other hand, are primarily based on cost benefit analysis of the consequences of the decision options. For both public and private decision making regarding the promotion or adoption of contract farming, it helps to follow principles of stakeholder and cost-benefit analyses in an attempt to enumerate the potential advantages and disadvantages of this mode of governance.

Advanteges for Farmers

Farmers find in contract farming a means to manage risks in production and marketing. For farmers, production and marketing risks are reduced because of the following reasons:

- inputs can be provided by agribusiness firms, thereby reducing the uncertainties associated with input availability, quality and costs. Failures in input markets are circumvented by such direct provision and the economies of scale allowed by the larger purchases of inputs by agribusiness firms can be passed on to farmers, via reduced costs.

Input quality and adequacy to the crop or livestock activity contracted is ensured and translated into higher productivity and higher returns. services, such as mechanisation and transportation, can be provided by agribusiness firms. In the absence of a contractual link, cost effective access to mechanised land preparation, sowing and harvest might be unfeasible to an individual farmer

— technological assistance can be offered by the contracting firm, favouring the production of higher valued, often riskier crops and livestock. Without technical assistance, the tendency is for farmers to avoid venturing into less known crop and livestock enterprises. A related benefit is the facilitation of the conversion path from subsistence to commercial farming.

— farm production and management skills are enhanced by the technical assistance provision, and spill-over effects might happen if farmers also engage in noncontracted crops and livestock enterprise activities. Such positive externalities of contracts have been observed in cases where subsistence crops are produced alongside cash crops grown under contractual arrangements.

— a market outlet is secured for the contracted production, such that the uncertainty and the transaction costs involved in the search for markets are reduced. Small-scale farmers in particular benefit from the reduction of marketing risks, as they often have more limited market access

— the uncertainty about sales price is often reduced, since contracts typically specify at the beginning of the growing cycle the prices to be paid at product delivery. Although price formulas used in some types of contracts may specify a relationship with uncertain, future prices prevailing in international or domestic markets, this is not the norm but rather an exception.

— with the reduction of product and market risks, income stability is favoured. If the contract has a long term continuation, farmers can benefit from the more foreseeable and stable income streams, especially via better planning of consumption and investment decisions

— access to credit is enhanced. Under a resource provision contract, working capital credit is typically supplied in kind, via input provision, by the contracting firm. The contract firms, or the banking system, may also offer investment credit for the acquisition of machinery, buildings, etc. These transactions are guaranteed by the commercial commitment between the borrower and the agribusiness firm. By the same token, credit access for both working and fixed capital is enhanced in the case of market specification contracts, because banks accept the contractual commitment as a sufficient guarantee for the granting of loans.

— by-products and residues from the contracted farming activity can be used in complementary farm enterprises. Poultry manure and sugar beet leaves are two examples that can illustrate this potential benefit. The former can be used by farmers in vegetable production for cash or subsistence products while the latter can be used as animal feed.

Disadvantages for Farmers

Even though farming risks can be reduced by contracting, the contracts themselves represent a risk source in the farm operation, thus leading to potential disadvantages for farmers. As we will see, most of these negative aspects come from the fact that the relationship between individual farmers and the agribusiness firm is uneven, the latter being often in a position to exercise power and noncompetitive conduct in the definition of the terms of the transactions. The potential disadvantages are:

— firms might renege on contractual terms if market circumstances change or if other conditions for opportunistic behaviour arise. Since prices specified in contracts are based on expectations about future market behaviour, substantial variations in the realisation of the expectations can lead firms to force renegotiation or to engage in contractual hold-up. Causes for such variations might be a bountiful supply depressing market prices, a change in government policy that allows imports of competitive products and a variation in exchange rates that significantly raises input prices, to name a few.

  A disguised form of hold-up is the rejection of products delivered, under pretext of non-conformity to quality regulations—firms might refuse to receive products as a strategy to transfer to farmers the financial losses arising from unexpected market turns. In the absence of effective enforcement mechanisms, there is little that a farmer can do to avoid the negative impact of contractual hold-ups.

— the dependency on a prescribed technology package makes farmers vulnerable to output and productivity manipulation by agribusiness firms. Manipulation, in this hypothesis, can be seen as an indirect, sophisticated means to control payments to farmers. An animal feed supplied by a firm, for instance, might be formulated so as to delay the growth cycle or decrease the daily weight gain, thus reducing productivity and the final prices paid to farmers. Technology control can thus be used as a conduit to shift market pressures from firms to producers.

— delivery schedules might be set by firms so as to influence prices paid to farmers. This strategy can happen when prices are rapidly changing and firms adjust the delivery schedule to benefit from the market volatility. It may also happen because of asset specificity related to timing of delivery. In the case of sugar cane, for instance, the delay might depress prices received if these are based on the degree of sucrose, which falls rapidly after harvest.

— firms might intentionally avoid transparency in the price determination mechanism of the contract, utilising complex formulas or quantity and quality measurements not well understood by farmers. Formula prices related to quality attributes such as fat content, somatic cell counts, sucrose content, bacterial counts and other criteria that require laboratory measurement lend themselves to fraud and manipulation. Quantity determination based on tare measurement is another example of contractual practice over which farmers can exercise little or no control.

— farmers lose flexibility in enterprise choice. Bound to a crop or livestock enterprise by a contract, farmers can not adjust production mixes so as to benefit from market opportunities

— long term contracts might lead to gradually decreasing real prices received by farmers. As asset specificity is built in the farm operation, firms might use this constraint as a way to establish and reinforce monopsonistic power and thus gradually impose lower prices on farmers.

— farmers may lose linkages with former transaction partners. Long-term business relationships with intermediaries, lenders and input providers might have been developed by farmers prior to engaging in a contractual relationship. These linkages are broken and can be difficult to re-establish if the event of exit from the contractual relationship.

— farmers may abandon traditional cultivation methods and products. Age old patterns of cultivation and animal rearing are disrupted by the innovations introduced in a contractual relationship. Patterns that emerged from the optimal utilisation of locally available resources might be irreversibly lost, as farmers become used to different technologies. Again, in the case of exit, it might be difficult to re-establish the traditional ways.

— the risks that are normally associated with see practices are increased. Intensified production of single agricultural crops, or the concentration of animal herds, increases the chances of diseases. Disease control requires the use of plant protection and veterinary inputs that might have undesirable environmental and human health effects.

— socio-cultural structures in the family or farming community might be disrupted. In societies where the division of labour and family responsibilities are shared by genders in distinctive ways, contracts can introduce threats to the cultural pattern. This can happen when farm labour is mostly performed by one gender, but cash payments by firms are made to the spouse. Social tensions might arise, in such cases.

— the risk of indebtedness grows. The downside of easy access to credit is the possibility to incur mounting debts. As farmers find fewer restrictions to access credit, they might find an incentive to borrow, even to finance consumption and other non-commercial needs.

— there is a tendency for the development of a gradual reliance on the contracting company for non-farm-related matters. With their dominant position, agribusiness firms might become a last resort benefactor in areas where public services are weak or non-existent. Firms may provide some social services and do small favours to farmers, such as the provision of free transportation rides or the distribution of promotional gifts (t-shirts, caps, pencils, etc.). The reliance on the firm to provide such services and favours weakens a farmer's bargaining position and reinforces the firm's ability to exercise monopsonistic conduct.

Advantages for Firms

The theoretical hypothesis is that agribusiness firms find in contract farming a strategy to minimise transaction costs, primarily the ones related to asset specificity and uncertainty. The main potential advantages are as follows:

— greater regularity of agricultural product supplies to the firm is ensured. With contracts, firms can schedule the deliveries so as to optimally utilise their processing capacity and / or distribution infrastructure. They can also better coordinate product delivery with the timing of the demands from their own clients.

— greater conformity to desirable product quality attributes and to safety standards is promoted. Since contracts specify quality attributes and since most also allow control of farming technology processes, firms are in a better position to meet consumer requirements and mandatory quality and safety standards.

— access to land is facilitated. Access to the large land areas needed by agribusiness enterprises may be precluded by legislation that caps farm sizes or exclude private companies from land ownership. Moreover, even where legislation is not restrictive, land can be costly to purchase. Contracting allows circumventing these limitations.

— input costs per unit are reduced. Economies of scale in purchasing can be attained by firms that acquire large quantities of farm inputs. These economies can stem from reasons such as the greater bargaining power of the firms, the reduced costs of bulk transportation, the sharing of fixed purchasing costs into a large number of purchased units and the possibility to access financing under more favourable conditions. Lower input costs imply lower production costs and thus permit that farmers maintain a given margin even when product prices are set at competitively lower levels.

— access to agricultural credit and eventual financial incentives and subsidies is facilitated. The reduction of risks in the firm's supply chain and the economies of scale associated with contracting operations are conditions that in principle increase a financing institution's willingness to lend. As earlier suggested, not only access is improved, but financing conditions tend to be more favourable to the firm. As to government incentives and subsidies, countries that promote contracting often use such instruments in their agribusiness development policies. They might be expressed in the form of tax breaks, foreign exchange quotas, profit repatriation flexibility, tariff reduction for imported inputs, etc.

— labour costs are reduced. Compared to full vertical integration, where the firm must hire and manage its own labour force, in contract farming these administrative responsibilities are transferred to farmers. Labour costs, including wages, social benefits, supervision and training, can represent a significant share of production costs in labour intensive farming operations. Farmers will use family labour and when hiring, they may not be bound to the same labour laws as agribusiness firms. Hence, labour costs tend to be lower under contracting schemes.

— expansion and contraction of production is facilitated. Without fixed assets in land or specialised housing for animals, for instance, agribusiness firms have greater flexibility to expand or reduce operations. This allows better adjustment to market developments.

— for high value, labour intensive agricultural enterprises, managerial efficiency in farming may be favoured. Under contracting schemes, there is evidence indicating that family farming units tend to achieve comparable or even better productivity, when compared with larger, commercially managed units. The main reasons are the invested commitment and the comparative advantage in micro-managing farming operations. An example is the production efficiency competitions among contracting farmers, regularly promoted by a poultry processor in south-eastern Brazil. In such comparisons, small farms were frequently in the top ranks.

Disadvantages for Firms

As for farmers, agribusiness firms incorporate new risk sources in their operations, when opting for contractual arrangements as a governance mode in their supply chains. These risks bring disadvantages for the use of contracts. Classic disadvantages are:

- *risk of contractual hold-ups.* Just as a firm may be prone to renege on contractual terms when market conditions change, a farmer may be compelled to sell all, or part of his or her production, to a third party, when prices are perceived to be higher outside the contractual bond. This is especially problematic where alternative markets for the crop or livestock grown under contracts are easily accessible and where contractual enforcement is weak. Contractual hold-ups by farmers are known to be the reason for the failure of contract farming in many such schemes.
- *transaction costs of dealing with large numbers of farmers are high.* A contracting firm will typically be linked to a sizeable number of farmers, scattered over wide regional areas. Managing a commercial relationship with a myriad of partners is a complex task, requiring investments in personnel, in controls

and in monitoring systems. The logistical costs tend also to be high when inputs must be distributed and production assembled by the contracting firm. For these reasons, some evidence in the literature of contract farming suggests that firms prefer to work with larger, rather than smaller farmers.

— *risk of misuse or deviation of supplied inputs and of final products.* In resource provision contracts, a known problem is the potential use of the distributed inputs in alternative crop and livestock enterprises. Farmers may use fertilisers in their subsistence crops or may feed domestic herds or flocks with feedstuffs provided for the contracted animal rearing. Farmers may also consume part of the production or, as we have already seen, simply sell to third parties. Firms tend to avoid this problem by closer monitoring and control and by relating prices paid to the efficiency of input use. In poultry growing contracts, formula prices take into account the feed conversion rate and the percentage mortality rate of the flock, among other performance indicators. If inputs or outputs are deviated, these figures should reflect it.

— *internalisation of support service costs.* In the absence of contracts, farmers seek support services with external providers, public or non public (private firms, NGO's etc). With contracts instead, extension services, transportation services, financing, quality monitoring and other services are transferred to the responsibility of the agribusiness firms. Such new costs must be internalised and reckoned in the decision to contract or not.

— *loss of flexibility to seek alternative supply sources.* Bound by a contract, a firm is precluded from sourcing from alternative suppliers when economic conditions would so advise. As previously commented, in situations of over supply in domestic markets or in cases of favourable movements of exchange rates that reduce import costs, contracting firms must still honour the contractual commitments and may lose competitiveness

— *risk of undermining the corporate image*. Firms with strong brand image are particularly affected by this potential risk. Contract farming is still seen in some circles as an exploitative relationship, where farmers become simply an outsourced labour force to rent-seeking companies. Moreover, conflicts with farmers may arise, or negative impacts of the contracted operations on the environment or in animal welfare, for instance, might exist. The association of a corporate image to these negative aspects is clearly undesirable. In brief, there are potential costs and benefits and there are less tangible rewards and shortcomings in contract farming, for all partners involved in it. The recent resurgence of contracting in agri-food systems suggests that the balance has been favourable to the participants. Successful contracting, nonetheless, demands some pre-conditions to be in place and these are reviewed in the next section.

Success Factors Contract Farming

As a preamble to the discussion of success factors for contract farming, a basic tenet might be enunciated as follows: contractual relationships will only be sustainable if partners perceive that they are better off by engaging in it. Both have to see the partnership as a source of gains—financial or otherwise—that could not be matched in the next best alternative. The corollary of this proposition is that contract farming will fail if parties do not develop mutual trust and reciprocal dependency rooted in the systems principle of synergy, whereby the whole becomes greater than the sum of its parts. As such, the critical factors for successful contract farming are primarily those that address the set of potential disadvantages that can countervail the sources of synergies.

A fundamental prerequisite is an appropriate enabling environment. No successful contracting scheme can exist or remain sustainable where the institutional and political setting is not conducive to it. The importance of this normative and policy framework can not be overemphasised—governments wishing to promote farmagribusiness linkages via contracts should start by assessing, streamlining and expunging the framework from restrictive features.

The normative framework must also deal explicitly with the issue of contractual hold-up. The obligations of parties involved in formal agreements on commercial transactions have to be enforceable, and legislation to that end should be in place. Most importantly, enforcement has to be operationally and financially feasible. If the judicial system is such that the cost of litigation becomes greater than its potential benefit, no legislation protecting the rights of contractual partners can be effectively enforced.

Since the minimisation of contractual hold-ups is considered as a critical success factor, strategies towards accomplishing this objective can be pursued by firms and farmers as complementary measures to legal enforcement. From the farmer's perspective, enhancement of bargaining power via collective action is a standard approach, for that matter. Through their associations, farmers can have stronger voices in negotiation processes and in this way better protect their interests. Firms, on the other hand, can reduce farmer defaults by group lending, by good communication and close monitoring, by the range and quality of services they offer and by treating defaulters strictly.

Providing credit and distributing inputs through groups reinforces peer pressure processes and in this way discourages non-compliance with the obligations individually assumed. Good communication fosters non-adversarial relationships between farmers and firms, whereas monitoring allows preventive actions when default is likely. Good quality and ample scope of services provided, of course, increase the opportunity cost of a farmer's default. Strict treatment of defaulters comprises the exclusion from the contractual scheme, asset seizure, charging of interest in payments due, or taking legal action—they are harsh, but nonetheless mostly effective measures that are ultimately beneficial to the group as a whole. Finally, a further way to reduce hold-up is to extend the contractual duration, as long term contracts in principle encourage investments in specific assets which, in turn, reduce the motivation for defaults.

The organisation of farmers in groups or other associative form is a strategy that can address two other success factors identified in this work. They are the need to countervail the unevenness in bargaining power and the need to reduce the transaction costs of dealing with multiple contracting parties.

In the first case, the concern with unfair conduct by contracting firms is justified by empirical evidence that imbalanced power in contractual relations can lead to noncompetitive behaviour by the dominant party. In the case of contracting agribusiness firms, this can be expressed, inter alia, by the imposition of low prices, by deductions of highly set input costs, by early termination of contracts, by the manipulation of quality attributes or by the design of biased contractual clauses.

Whereas farmers' group action per se does not impede such practices, it does reduce the scope for their imposition. Besides with their strengthened power for financial negotiations, associations are better positioned to find support from government, NGO's or private advisors in order to monitor compliance with contractual clauses, double check product quality measurements, mediate litigations and provide information on prices and costs.

13

Advantages and Problems of Contract Farming

The advantages, disadvantages and problems arising from contract farming will vary according to the physical, social and market environments. More specifically, the distribution of risks will depend on such factors as the nature of the markets for both the raw material and the processed product, the availability of alternative earning opportunities for farmers, and the extent to which relevant technical information is provided to the contracted farmers. These factors are likely to change over time, as will the distribution of risks.

Advantages for Farmers

The prime advantage of a contractual agreement for farmers is that the sponsor will normally undertake to purchase all produce grown, within specified quality and quantity parameters. Contracts can also provide farmers with access to a wide range of managerial, technical and extension services that otherwise may be unobtainable. Farmers can use the contract agreement as collateral to arrange credit with a commercial bank in order to fund inputs. Thus, the main potential advantages for farmers are:

— provision of inputs and production services;
— access to credit;
— introduction of appropriate technology;
— skill transfer;
— guaranteed and fixed pricing structures; and
— access to reliable markets.

Access to Inputs and Production Services

Many contractual arrangements involve considerable production support in addition to the supply of basic inputs such as seed and fertiliser. Sponsors may also provide land preparation, field cultivation and harvesting as well as free training and extension. This is primarily to ensure that proper crop husbandry practices are followed in order to achieve projected yields and required qualities. There is, however, a danger that such arrangements may lead to the farmer being little more than a labourer on his or her own land.

It is often difficult for small-scale farmers outside the contract-farming context to gain access to inputs. In Africa, in particular, fertiliser distribution arrangements have been disrupted by structural adjustment measures, with the private sector having yet to fill adequately the void created by the closure of parastatal agencies. In many countries a vicious circle has developed whereby the low demand for inputs provides no incentive for the development of commercial distribution networks and this, in turn, further adversely affects input availability and use. Contract farming can help to overcome many of these problems through bulk ordering by management.

Access to Credit

The majority of smallholder producers experience

difficulties in obtaining credit for production inputs. With the collapse or restructuring of many agricultural development banks and the closure of many export crop marketing boards (particularly in Africa), which in the past supplied farmers with inputs on credit, difficulties have increased rather than decreased.

Contract farming usually allows farmers access to some form of credit to finance production inputs. In most cases it is the sponsors who advance credit through their managers. However, arrangements can be made with commercial banks or government agencies through crop liens that are guaranteed by the sponsor, i.e. the contract serves as collateral. When substantial investments are required of farmers, such as packing or grading sheds, tobacco barns or heavy machinery, banks will not normally advance credit without guarantees from the sponsor.

The tendency of certain farmers to abuse credit arrangements by selling crops to buyers other than the sponsor (extra-contractual marketing), or by diverting inputs supplied by management to other purposes, has caused some sponsors to reconsider supplying most inputs, opting instead to provide only seeds and essential agrochemicals. The policies and conditions that control advances are normally described in attachments to contracts.

Adoption of New Technology

New techniques are often required to upgrade agricultural commodities for markets that demand high quality standards. New production techniques are often necessary to increase productivity as well as to ensure that the commodity meets market demands. However, small-scale farmers are frequently reluctant to adopt new technologies because of the possible risks and costs involved. They are more likely to accept new practices when they can rely on external resources for material and technological inputs. Nevertheless, the introduction of new technology will not be successful unless it is initiated within a well-managed and structured farming operation. Private agribusiness will usually offer technology more diligently than government agricultural extension services because it has a direct economic interest in improving farmers' production. Most of the larger sponsors prefer to provide their own extension rather than rely on government services.

Skill Transfer

The skills the farmer learns through contract farming may include record keeping, the efficient use of farm resources, improved methods of applying chemicals and fertilisers, a knowledge of the importance of quality and the characteristics and demands of export markets. Farmers can gain experience in carrying out field activities following a strict timetable imposed by the extension service. In addition, spillover effects from contract farming activities could lead to investment in market infrastructure and human capital, thus improving the productivity of other farm activities. Farmers often apply techniques introduced by management (ridging, fertilising, transplanting, pest control, etc.) to other cash and subsistence crops.

Guaranteed and fixed pricing structures

The returns farmers receive for their crops on the open market depend on the prevailing market prices as well as on their ability to negotiate with buyers. This can create considerable uncertainty which, to a certain extent, contract farming can overcome. Frequently, sponsors indicate in advance the price(s) to be paid and these are specified in the agreement. On the other hand, some contracts are not based on fixed prices but are related to the market prices at the time of delivery. In these instances, the contracted farmer is clearly dependent on market volatility.

Access to reliable markets

Small-scale farmers are often constrained in what they can produce by limited marketing opportunities, which often makes diversification into new crops very difficult. Farmers will not cultivate unless they know they can sell their crop, and traders or processors will not invest in ventures unless they are assured that the required commodities can be consistently produced. Contract farming offers a potential solution to this situation by providing market guarantees to the farmers and assuring supply to the purchasers.

Even where there are existing outlets for the same crops, contract farming can offer significant advantages to farmers. They do not have to search for and negotiate with local and international buyers, and project sponsors usually organise transport for their crops, normally from the farmgate.

Disadvantages for Farmers

For farmers, the potential problems associated with contract farming include:

— increased risk;

— unsuitable technology and crop incompatibility;

— manipulation of quotas and quality specifications;

— corruption;

— domination by monopolies; and
— indebtedness and overreliance on advances.

These potential problems can usually be minimised by efficient management that consults frequently with farmers and closely monitors field operations.

Increased Risk

Farmers entering new contract farming ventures should be prepared to balance the prospect of higher returns with the possibility of greater risk. Such risk is more likely when the agribusiness venture is introducing a new crop to the area. There may be production risks, particularly where prior field tests are inadequate, resulting in lower-than-expected yields for the farmers. Market risks may occur when the company's forecasts of market size or price levels are not accurate. Considerable problems can result if farmers perceive that the company is unwilling to share any of the risk, even if partly responsible for the losses. In Thailand, for example, a company that contracted farmers to rear chickens charged a levy on farmers' incomes in order to offset the possibility of a high chicken mortality rate. This was much resented by the farmers, as they believed that the poor quality of the day-old chicks supplied by the company was one reason for the problem.

Unsuitable Technology and Crop Incompatibility

The introduction of a new crop to be grown under conditions rigorously controlled by the sponsor can cause disruption to the existing farming system. For example, the managers may identify land traditionally reserved for food crops as the most suitable for the contracted crop. Harvesting of the contracted crop may fall at the same time as the harvesting of food crops, thus causing competition for scarce labour resources. Particular problems may be experienced when contract farming is related to resettlement programmes. In Papua New Guinea, for example, people from the Highlands were resettled in coastal areas to grow oil palm and rubber. This required the farmers, who were traditionally sweet potato eaters, to learn cultivation techniques for new food crops and to adapt their dietary practices accordingly.

Two factors should be considered before innovations are introduced to any agricultural environment. The first is the possible adverse effect on the social life of the community. When tobacco growers in Fiji were encouraged to cure tobacco themselves rather than sell it in the fresh green form, it was found that they were unable to handle the highly technical curing operation with any degree of continuity. This was attributed to intermittent social commitments and customary obligations that overrode contractual responsibilities and eventually resulted in the cancellation of their contracts.

The second factor is the practicality of introducing innovations or adaptations. The introduction of sophisticated machines (e.g. for transplanting) may result in a loss of local employment and overcapitalisation of the contracted farmer. Furthermore, in field activities such as transplanting and weed control, mechanical methods often produce less effective results than do traditional cultivation methods. Field extension services must always ensure that the contracted crop fits in with the farmer's total cropping regime, particularly in the areas of pest control and field rotation practices.

Manipulation of Quotas and Quality Specifications

Inefficient management can lead to production exceeding

original targets. For example, failures of field staff to measure fields following transplanting can result in gross overplanting. Sponsors may have unrealistic expectations of the market for their product or the market may collapse unexpectedly owing to transport problems, civil unrest, change in government policy or the arrival of a competitor. Such occurrences can lead managers to reduce farmers' quotas. Few contracts specify penalties in such circumstances. In some situations management may be tempted to manipulate quality standards in order to reduce purchases while appearing to honour the contract. Such practices will cause sponsor-farmer confrontation, especially if farmers have no method to dispute grading irregularities. All contract farming ventures should have forums where farmers can raise concerns and grievances relating to such issues.

Corruption

Problems occur when staff responsible for issuing contracts and buying crops exploit their position. Such practices result in a collapse of trust and communication between the contracted parties and soon undermine any contract. Management needs to ensure that corruption in any form does not occur. On a larger scale, the sponsors can themselves be dishonest or corrupt. Governments have sometimes fallen victim to dubious or "fly-by-night" companies who have seen the opportunity for a quick profit. Techniques could include charging excessive fees to manage a government-owned venture or persuading the government and other investors to set up a new contract farming company and then sell that company overpriced and poor quality processing equipment. In such cases farmers who make investments in production and primary processing facilities run the risk of losing everything.

Domination by Monopolies

The monopoly of a single crop by a sponsor can have a negative effect. Allowing only one purchaser encourages monopolistic tendencies, particularly where farmers are locked into a fairly sizeable investment, such as with tree crops, and cannot easily change to other crops. On the other hand, large-scale investments, such as for nucleus estates, often require a monopoly in order to be viable. In order to protect farmers when there is only a single buyer for one commodity, the government should have some role in determining the prices paid.

Drucker suggests that privately managed monopolies under public regulation are preferable to non-regulated private or public monopolies. The greatest abuses do tend to occur when there are public monopolies, where buying prices are set by the government, or where farmers have made long-term investments in perennial crops. In 1999 the Kenya Tea Development Authority experienced serious unrest amongst its growers, reportedly because of the Authority's inefficient extension services and alleged "manipulation" of farmers. There was also discontent in Kenya among sugar farmers because the price set by the government did not change between 1997 and 1999.

Indebtedness and Overreliance on Advances

One of the major attractions of contract farming for farmers is the availability of credit provided either directly by the company or through a third party. However, farmers can face considerable indebtedness if they are confronted with production problems, if the company provides poor technical advice, if there are significant changes in market conditions, or if the company fails to honour the contract. This is of particular concern with long-term investments, either for tree crops or for on-farm processing facilities. If advances are uncontrolled, the indebtedness of farmers can increase to uneconomic levels. In one venture "compassionate" advances for school fees, weddings and even alimony resulted in many farmers receiving no payments at the end of the season. Dropout rates for farmers in that particular project were high, as they thought contract farming did not pay.

Advantages for Sponsors

Companies and government agencies have a number of options to obtain raw materials for their processing and marketing activities. The benefits of contract farming are best examined in the light of the other alternatives, namely spot-market purchases and large-scale estates. The main potential advantages for sponsors can be seen as:

— political acceptability;
— overcoming land constraints;
— production reliability and shared risk;
— quality consistency; and
— promotion of farm inputs.

Political Acceptability

It can be more politically expedient for a sponsor to involve smallholder farmers in production rather than to operate plantations. Many governments are reluctant to have large plantations and some are actively involved in closing down such estates and redistributing their land. Contract farming, particularly when the farmer is not a tenant of the sponsor, is less likely to be subject to political criticism. As a result of the restructuring of their economies, many African governments have promoted contract farming as an alternative to private, corporate and state-owned plantations. In Zimbabwe, for example, contract farming is actively encouraged, particularly in the sugar-cane, tea and cotton industries.

In recent years many countries have seen a move away from the plantation system of production to one where smaller-scale farmers grow crops under contract for processing and/or marketing. In Central America, for example, multinational corporations have moved from banana plantation production to purchasing bananas grown by contracted farmers, with the corporations providing technical advice and marketing services. This trend is also found in the international tobacco industry; smallholder tobacco production through contract farming has replaced estates in several countries. Similar changes have occurred with other crops. In Kenya, the tea industry, originally founded on the plantation model, now provides extension services and inputs to tens of thousands of contracted farmers.

The decision to choose contract farming does not make a company totally immune from criticism. For example, the considerable opposition to the role of multinational corporations in India in the late 1990s had a negative effect on investment in contract farming by foreign agribusiness corporations.

Overcoming land Constraints

Most of the world's plantations were established in the colonial era when land was relatively plentiful and the colonial powers had few scruples about either simply annexing it or paying landowners minimal compensation. That is, fortunately, no longer the situation. Most large tracts of suitable land are now either traditionally owned, costly to purchase or unavailable for commercial development. Moreover, even if it were possible for companies to purchase land at an affordable price, it would rarely be possible to purchase large enough parcels of land to offer the necessary economies of scale achieved by estate agriculture. Contract farming, therefore, offers access to crop production from land that would not otherwise be available to a company, with the additional advantage that it does not have to purchase it.

Although it may be considered that plantation agriculture on a large scale is generally more cost-effective than small-scale production, that is not always the case. Estate production involves both direct labour costs and indirect costs of labour in terms of hiring, training and supervising. It is often necessary to provide accommodation and meals for estate workers. As noted above, land can be very expensive and difficult to obtain, thus contract farming can often be competitive, particularly for crops where large-scale economies of scale are difficult to achieve. As already noted, experience in some developing countries indicates that plantation models of crop production can evolve successfully into cost-effective smallholder contract farming ventures.

Production Reliability and Shared Risk

The failure to supply agreed contracts could seriously jeopardise future sales. Plantation agriculture and contract farming both offer reasonable supply reliability. Sponsors of contract farming, even with the best management, always run the risk that farmers will fail to honour agreements. On the other hand, plantation agriculture always runs the risk of labour disputes. In the case of horticultural production some companies do prefer estate rather than contracted production. In

Gambia and Ghana, for example, a number of crops are grown under the estate model, as are strawberries and flowers in Kenya.

Working with contracted farmers enables sponsors to share the risk of production failure due to poor weather, disease, etc. The farmer takes the risk of loss of production while the company absorbs losses associated with reduced or non-existent throughput for the processing facility. Where production problems are widespread and no fault of the farmers, sponsors will often defer repayment of production advances to the following season.

Both estate and contract farming methods of obtaining raw materials are considerably more reliable than making purchases on the open market. The open market is rarely an acceptable option for organisations that have significant assets tied up in processing facilities and need to have guaranteed quantities of raw material to justify their investment. For example, it is hardly ever an acceptable option for companies who make regular shipments of horticultural produce to supermarkets and for export. Companies must ensure that crops are harvested and sold on a carefully scheduled and consistent basis: a factor that is normally assured under a well-directed contract farming scheme.

Quality Consistency

Markets for fresh and processed agricultural produce require consistent quality standards. Moreover, these markets are moving increasingly to a situation where the supplier must also conform to regulatory controls regarding production techniques, particularly the use of pesticides. For fresh produce there is an growing requirement for "traceability", i.e. suppliers to major markets increasingly need to be confident of identifying the source of production if problems related to food safety arise. Both estate and contracted crop production require close supervision to control and maintain product quality, especially when farmers are unfamiliar with new harvesting and grading methods. Often, large numbers of crops within a single project have to be transplanted, harvested and purchased in a uniform manner so as to achieve product consistency.

Distinct varieties of produce in the desired quality and quantities are often not available on the open market. For example, a multinational that invested in the Indian State of Punjab found that the local varieties of tomatoes were unsuitable for processing into paste or ketchup. This was one of the factors that made it decide to go into contract farming.

Agribusinesses producing for markets demanding high quality standards, such as fruits and vegetables for export, often find that small-scale farmers and their families are more likely to produce high-quality products than farmers who must supervise hired labour. Also contract farming makes quarantine controls more manageable. It is easier for quarantine authorities to inspect a limited number of exporters of a single commodity, who closely supervise farmers, than to inspect hundreds, or sometimes thousands, of individual producers selling through open markets. Much of the production of "organic" foods is being done on contract, as an integrated operation facilitates a clear crop identity from farmer to retailer. In some highly sophisticated operations, containers are now being loaded on the farm for direct delivery to the supermarket.

Promotion of Farm Inputs

An example of an unusual but, nevertheless, interesting

benefit for sponsors comes from the Philippines. A feed milling company experienced difficulties in marketing its feed, which was more expensive than that produced by competing companies. To solve this problem it developed rearing schemes for pigs and poultry under contract in order to provide a market outlet for its feeds and to demonstrate their performance to other farmers living near the contracted farmers.

Disadvantages for Sponsors

The main disadvantages faced by contract farming developers are:

— land availability constraints;

— social and cultural constraints;

— farmer discontent;

— extra-contractual marketing; and

— input diversion.

Land Availability Constraints

Farmers must have suitable land on which to cultivate their contracted crops. Problems can arise when farmers have minimal or no security of tenure as there is a danger of the sponsor's investment being wasted as a result of farmer-landlord disputes. Difficulties are also common when sponsors lease land to farmers. Such arrangements normally have eviction clauses included as part of the conditions. In Gambia, land rights are determined not only by gender but also by the historic manner of land use. When international donor organisations insisted on having a legal titleholder for contracted crops, resistance to giving women formal titles to land was shown by male household heads. The objection was based on the fear of permanent land alienation that could occur as the result of matrimonial disputes.

Some contract farming ventures are dominated by customary land usage arrangements negotiated by landless farmers with traditional landowners. While such a situation allows the poorest cultivator to take part in contract farming ventures, discrete management measures need to be applied to ensure that landless farmers are not exploited by their landlords. Before entering into contracts, the sponsor must ensure that access to land is secured, at least for the term of the agreement.

Social and Cultural Constraints

Problems can arise when management chooses farmers who are unable to comply with strict timetables and regulations because of social obligations. Promoting agriculture through contracts is also a cultural issue. In communities where custom and tradition play an important role, difficulties may arise when farming innovations are introduced. Before introducing new cropping schedules, sponsors must consider the social attitudes and the traditional farming practices of the community and assess how a new crop could be introduced. Customary beliefs and religious issues are also important factors. For example, Easter for some Christians is an inappropriate time for sowing vegetable crops. Harvesting activities should not be programmed to take place during festivals, and failure to accommodate such traditions will result in negative farmer reaction. It must also be recognised that farmers require time to adjust to new practices.

Farmer Discontent

A number of situations can lead to farmer dissatisfaction. Discriminatory buying, late payments, inefficient extension services, poor agronomic advice, unreliable transportation for crops, a mid-season change in pricing or management's rudeness to farmers will all normally generate dissent. If not readily addressed, such circumstances will cause hostility towards the sponsors that may result in farmers withdrawing from projects. This emphasises the importance of good management to the success of contract farming.

Extra-contractual Marketing

The sale of produce by farmers to a third party, outside the conditions of a contract, can be a major problem. Extra-contractual sales are always possible and are not easily controlled when an alternative market exists. For example, a farmer cooperative in Croatia bought cucumbers, red peppers and aubergines on contract. The cooperative's advances to the farmers included all necessary production inputs. Unfortunately members often sold their vegetables to traders at higher prices than the cooperative had contracted. The outside buyers offered cash to farmers as opposed to the prolonged and difficult collection of payments negotiated through the cooperative. Sponsors themselves can sometimes be a cause of extra-contractual practices. In Colombia, a company purchased passion fruit from a competitor's growers when production shortfalls occurred. A similar situation was also experienced in Indonesia where a number of sponsors competed for quality tobacco by surreptitious means. This led to a "tobacco war" between various sponsors that eventually forced the local provincial government to intervene.

In another case, a tobacco project diversified into off-season maize to provide farmers with additional income. In the first season some farmers sold their maize crops to traders for cash. Over 60 percent of the first season's maize crop was estimated to have been sold outside the agreement. The repayment of loans advanced for inputs was thereby circumvented, making the diversification venture uneconomical for the sponsor. The sponsor imposed strict penalties the following year as part of the maize registration formula. If the farmers were found to be selling their maize outside the agreement, their highly profitable tobacco agreement was cancelled.

Where there are several companies working with the same crop, they could collaborate by establishing a register of contracted farmers. Managers must be aware of produce being sold outside the project and also be aware of produce from outside being channelled into the buying system. This occurs when non-contracted farmers take advantage of higher prices paid by an established sponsor. Non-contracted crops are filtered into the buying system by outside farmers through friends and family who have crop contracts. Such practices make it difficult for the sponsor to regulate production targets, chemical residues and other quality aspects.

Input Diversion

A frequent problem is that farmers are tempted to use inputs supplied under contract for purposes other than those for which they were intended. They may choose to use the inputs on their other cash and subsistence crops or even to sell them. Clearly this is not acceptable to the sponsor, as the contracted crop's yields will be reduced and the quality affected. Steps to overcome such problems include improved monitoring by extension staff, farmer training and the issuing of realistic quantities of inputs. However, the knowledge that a contract has the advantages of technical inputs, cash advances and a guaranteed market usually makes the majority of farmers conform to the agreement.

14

Successful Contract Farming

The primary precondition for any investment in contract farming is profitable. Having identified a potentially profitable market the sponsor can then move on to assess whether that market can be profitably supplied by contracted farmers in a particular location of a particular country. This involves an assessment of the social and physical environment of the proposed contract area as well as the potential support likely to be provided by the government.

Profitable Market

Profit for the Sponsor

The sponsor's decision to invest in a particular market must be based initially on the knowledge that, subject to certain conditions, it will be profitable. However, contract farming is then just one of a number of solutions to a commercial market opportunity. A market must have the capacity to remain profitable in the longer term. In the case of tree crops, for example, prices tend to be cyclical. An analysis of economic viability carried out when prices are high would produce very different results than those obtained at the bottom of the price curve. A "sensitivity analysis" is thus required to ensure that production can be carried out profitably even when prices are low.

The exporting of horticultural produce to the markets of Western Europe, Japan and the United States is very competitive. Subject to guarantees regarding quality and supply, importers purchase produce on the basis of price. A supplier, through contact farming or otherwise, can lose markets overnight if quality standards and deliveries become unpredictable and inconsistent. Companies considering high-value horticultural exports also need to be certain that they can meet existing quality standards and likely future requirements.

Profit for the Farmer

If either the sponsors or their contracted farmers fail to achieve consistent and attractive financial benefits a venture will collapse. A further precondition, therefore, is that the sponsor needs to be sure that farmers will obtain higher net incomes from entering into a contract than they could from alternative activities with the same, or less, risk. Sponsors should calculate realistic yields in order to forecast whether production by farmers can be profitable at prices the sponsors are able to pay. These estimates should be based on the experience of farmers in the chosen area, their historical production data, soil fertility and, sometimes, field trials. Once estimates are compiled and production costs known, the sponsors are in a sounder position to calculate a realistic pricing structure that is mutually profitable. Guaranteed, regular and attractive incomes should encourage farmers to make a long-term commitment.

Sponsors should be aware that yield results from research plots are normally far higher than results from farmers' fields. Agronomists in Indonesia noted that soybeans grown at research stations produced yields more than twice those achieved by small-scale farmers. Experienced managers of contract farming projects usually estimate yields based on the mean production over the previous three to five years. As new technologies are introduced and farm management improves the mean yield increases over time. When a new crop is introduced the yield estimates are based on historic knowledge of the crop grown in similar environments and on the results of field trials.

Physical and Social Environments

The main factors affecting the success of all agribusiness ventures are:

- — the physical environment;
- — utilities and communications;
- — land availability and tenure;
- — input availability;
- — social considerations.

Physical Environment

The success of any agricultural investment requires that two multidimensional preconditions be met. Firstly, the general suitability of the topography, climate, soil fertility and water availability. Secondly, the suitability of the physical environment for the specific plant genotype or animal for which there is a market demand. The extent to which all these factors interact determines production yields, quality and profitability.

Utilities and Communications

A major precondition for agricultural investment in rural

areas is the existence of an adequate communication system that includes roads, transport, telephones and other telecommunication services. Reliable power and water supplies are particularly vital for agro-processing and exporting of fresh produce. The availability of suitable educational and medical services is also important for those who participate in contract farming, whether they be direct employees of the sponsor or the farmers themselves.

Sponsors will need to be assured that farm produce can be easily transported and that inputs can be delivered to their farmers. While major road infrastructure may be adequate, approach (or feeder) roads to farms may not. This is particularly important in the case of perishable crops that need to be processed soon after harvest (e.g. tea, oil palm and sugar) or stored in a suitable environment (e.g. cut flowers). Where local transport access is inadequate, sponsors must decide whether the problems can be resolved or whether alternate areas should be selected. Sometimes farmer groups are given the responsibility for ensuring that company transporters can reach the fields. Before the start of any project, the sponsor, farmers and local government agencies must agree on who will ensure access to and maintain feeder roads. In Kenya, the sugar companies' agreement with farmers stipulated that the companies had the right to construct feeder roads on the farmers' lands. This inevitably caused resentment among the landowners.

A precondition for the export of horticultural crops under contract is the availability of regular airfreight schedules; fresh vegetables and cut flowers depend on adequate cargo space to international markets. Unless quantities are large enough to justify chartering planes, the exporters will be dependent on space being available on commercial flights. The number of commercial flights depends on the number of passengers wanting to fly, and this can fluctuate rapidly. Several countries that have experienced coups or social disturbances have seen their tourism industries collapse overnight. This, in turn, has led to flight cancellations and the loss of markets for the exporters.

Land Availability and Tenure

Contract farming can involve a wide diversity of land ownership and tenure arrangements. Farmers under contract must have unrestricted access to land on which to plant their crops. There must be an awareness and understanding on the part of management of how farmers gain access to land for cultivation and for that access to be acceptable within the framework of the contract.

In the majority of projects, sponsors contract directly with farmers who either own land or have customary land rights within a communal landowning system. However, within a single project there can be numerous variations of land tenure, including freehold title, formal lease of state land, leases from the sponsor's own estate and informal seasonal arrangements with landlords. Even if tenure is on an extralegal, customary and seasonal basis, short-term contracted crops such as maize, tobacco and all table vegetables can often be accommodated.

Despite the occasionally flexible nature of customary land tenure, the dominant factor now controlling land tenure under contract farming is the rent demanded by the landowner. In one venture land rents were dependent on the whim of the respective landlords. This resulted in a wide variation of charges, influenced by the nature of individual farmer-landlord relationships. Some of the land rents were relatively low, many reasonable and some grossly inflated. In such cases interventions by sponsors may be necessary to negotiate standard rents on behalf of all farmers.

Input Availability

In most contract farming ventures the sponsors recommend, procure and distribute many or all of the material inputs. Sponsors need to be assured that they will be able to organise the supply of all necessary inputs for the farmers and for their own processing needs. All inputs should be identified and ordered well in advance, either from local sources or from overseas. Contract farming ventures call for varying levels of inputs depending on the nature of the crop and the degree of the farmers' sophistication. For crops such as Virginia flue-cured tobacco, farmers require a multitude of structural and material inputs that include curing barns, grading sheds, fuel, fertiliser, imported seed, pesticides and cultivation advances. Failure to have ready access to these can cause serious disruption to the production chain and can result in serious financial losses for all parties. Similarly, the failure of managers to supply feed on time to poultry and pig rearers can have major consequences for the farmers.

Social Considerations

Many rural communities are wary of modern agribusiness and strongly influenced by traditional practices. Conventional societies are normally more conservative in their ambitions and material needs. There are often great disparities in cultural attitudes towards work. Before beginning a venture, managers need to develop an understanding of the cultural attitudes of those with whom they are working. They must also be particularly aware of the possibility of disputes when there is more than one cultural group working on the contract.

There is always the possibility that the economic success of a contract farming venture could, in fact, have social repercussions that jeopardise its long-term success. This may occur, for example, because the opportunity to participate is limited to a certain number of farmers. If farmers are chosen on the basis of the size of their farms and resources, contract farming may widen pre-existing economic disparities and lead to resentment on the part of those excluded. In India there is concern that contract farming has led to a reversal of previous tenancy arrangements, with small-scale farmers now renting out land to large-scale farmers who have contracts.

Government Support

Governments have to play an important role if contract farming is to be successful. A relevant legal framework and an efficient legal system are preconditions. Moreover, governments can do much to foster success by developing linkages between investors and farmers and can play an important role in protecting farmers by ensuring the financial and managerial reliability of potential sponsors. The role of national governments and their local agencies can be divided into:

— the enabling and regulatory role; and

— the developmental role.

Enabling and Regulatory Role

Contract farming depends on either legal or informal agreements between the contracting parties. These, in turn, have to be backed up by appropriate laws and an efficient legal system. Relevant laws can be grouped into three categories: enabling functions, economic regulatory functions and constraining functions. In the context of contract farming the enabling aspect of the law is perhaps the most important. Laws of contract, in particular, allow the evolution of commercial transactions beyond direct barter exchanges. Legal mechanisms for granting a group of individuals recognition as a legal entity have also been central to the development of commerce. A classic example is the limited-liability company. However, in the context of contract farming, a sponsor entering into agreement with a cooperative also needs to be assured that the cooperative is on a sound legal footing.

Governments need to be aware of the implications of all laws and policy decisions on agribusiness development and how those policies influence contract farming. In the Philippines, for example, fast-food chains had been importing frozen french fries. Although that particular variety of potato could be grown in the Philippines, the Government had imposed import restrictions on seed potatoes, resulting in the unavailability of the required variety. Approaches to the Government by the companies eventually resulted in the ban being lifted and this permitted the establishment of two contract farming ventures to supply the rapidly growing fast-food industry. Thus a simple policy reform ultimately benefited the sponsors and a large number of small-scale farmers.

While it may not be considered a precondition it is desirable that governments play an arbitration or dispute resolution role. For example, the Government of Malawi established dispute resolution guidelines for agricultural contracts and offered the services of the Ministry of Labour to mediate. Likewise, in many large-scale, sugar-producing countries there are statutory bodies that act as arbitrators between sugar-cane growers and the sugar mills. In Canada, thousands of potato growers under contract with a single buyer negotiate prices and contract terms through the offices of the New Brunswick Potato Agency. It is compulsory that all potato farmers join the Agency.

Other government enabling activities to sustain contract farming may include:

- — Provision for training in technological and managerial skills at all levels, if sponsors do not provide those services.
- — Initiation and facilitation of research studies into the product under contract, in collaboration and consultation with the sponsors. State research institutes can particularly benefit smaller ventures, especially those managed by individual developers who cannot sustain their own plant breeding programmes, etc.
- — Provision of agricultural extension services to ventures that do not employ their own field staff. Small-scale developers cannot afford the luxury of their own extension service and thus need to make use of government services.

At the national level, it is a precondition that specialised services are available to provide institutional support to production, processing and marketing. Government services, such as quarantine controls, plant pathology clinics and research stations are important for contract farming. Such services are particularly necessary for companies that invest in high-value crops for export or in organic farming.

In most countries there is no legislation that specifically regulates contract farming. If legislation is introduced it should ideally be based on the industry's ability to regulate itself. However, governments have sometimes attempted to overregulate. This is often done when the sponsor is a parastatal or other government agency. For example, legislation in Kenya authorised the parastatal sponsor of contract tea farming to issue licences to farmers on rigid conditions. These governed aspects such as authority to uproot tea bushes, pest and disease controls, unauthorised planting of tea, failure to cultivate in the approved manner, and the right of the parastatal to grant or refuse a licence to plant tea. Although regulations such as these may have done the opposite, it can be argued that governments should enact legislation to protect farmers as the weaker of the contracting parties. This is particularly the case where the farmers involved are tenants of the sponsors and have little security.

Businessmen, particularly those involved with exports, frequently complain about the red tape and the costs involved with complying with excessive bureaucratic regulations and procedures. A simplification of official documentation, for example, could have a positive impact on the outlook of potential investors.

Developmental Role

As contract farming grows in importance governments should perhaps reallocate development resources towards its promotion. For example, the Philippines Government, with assistance from an FAO project, promoted contract farming for small-scale farmers who were allocated land under the agrarian reform programme. A major feature of this was a "market matching" exercise. This involved organising forums where agribusiness entrepreneurs could meet farmers' representatives to discuss their requirements. The forums were followed by more detailed discussions between individual sponsors and individual cooperatives or farmer organisations. By 2000 at least 27 companies had established contractual relationships with farmers as a result of the programme. Other activities carried out by the Department of Agrarian Reform included dissemination of market information, highlighting the products for which there was a commercial demand that could be satisfied through contract farming operations. The Department also agreed to act as arbitrator in the case of disputes.

Another example of promotion of contract farming comes from India where the regional office of a government-owned bank organised a meeting of bankers, agribusiness executives and the government extension service in order to explore possibilities of creating market linkages for agricultural products. This led to a major poultry producer contracting 2 200 farmers in 164 villages to grow maize and soybeans for feed purposes. Finance is provided by the banks, with a tripartite agreement being signed by farmers, the company and the banks.

Where contracted farmers are organised into cooperatives or groups, governments can play an important role by carrying out activities to strengthen the managerial skills of these organisations. Although the performance of agricultural cooperatives in developing countries has been marginal at best, improving a cooperative's managerial capability should, in theory, greatly enhance its business performance, the transfer of technology to farmers and its marketing skills.

The government has a role to play in ensuring that companies proposing to invest in contract farming are bona fide and are planning long-term partnership arrangements with farmers, rather than short-term operations which may leave farmers with considerable debts. Sponsors must have demonstrated financial strength, proven managerial competence and technological experience. Before promoting and launching projects, sponsors should create a suitable management and administrative structure and the purchase or lease of land for offices, processing facilities and transport needs should be organised in advance. Some projects may involve considerable capital investment and elaborate infrastructure such as that required for sugar milling, tobacco processing and vegetable canning.

Bibliography

Anon, *Contract Farming Ventures in India: A Few Successful Cases*, Spice, Vol1. No. 4; Ministry of Agriculture, India, March 2003

Chen, K.; Shepherd A. and Da Silva. C., "Changes in Food Retailing in Asia: Implications of Supermarket Procurement Practices for Farmers and Traditional Marketing Systems", *FAO AGSF Occasional Paper* No. 8, 2005.

Coulter, J., Goodland, A., Tallontire, A. and Stringfellow, R., "Marrying Farmer Cooperation and Contract Farming for Agricultural Service Provision in Sub-Saharan Africa", Natural Resources Institute, 2000.

Eaton, C. and Shepherd. A., "Contract Farming: Partnerships for Growth", *FAO Agricultural Services Bulletin* 145, Rome, 2001.

Guo, H. Jolly, R. and Shu, J., "Contract Farming in China: Supply Chain or Ball and Chain?", *Paper presented in the Minnesota International Development Conference*, University of Minnesota, April 2005

Glover, D., "Contract Farming and Commercialization of Agriculture in Developing Countries", In: von Braun, J. and Kennedy, E. (Eds.). *Agricultural Commercialization, Economic Development and Nutrition*, Johns Hopkins, 1994

Hobbs, J., "A Transaction Costs Approach to Supply Chain Management", *Supply Chain Management*, 1996.

Huddleston, P., "Old Crop, New Crop: Contrcating Out Oil Palm Production in Ghana", *School of Earth and Geographical Sciences*, The University of Western Australia, 2005.

Key, N. and Runsten, D., "Contract Farming, Smallholders, and Rural Development in Latin America: The Organization of Agroprocessing Firms and the Scale of Outgrower Production", *World Development*, 1999.

Kirsten, J. and Sartorius, K.l., "Linking Agribusiness and Small Farmers in Developing Countires: Is There a New Role for Contract Farming?", Development Southern Africa 19 (4) October 2002

Key, Nigel, Nelson, K., Southard, L., "Contracts, Markets and Prices: Organizing the Production and Use of Agricultural Commodities", *Agricultural Economic Report* Number 837, USDA, November 2004.

Minot, N., "Contract Farming and its Effect on Small Farmers in Less Developed Countries", *Working Paper* No. 31. Department of Agricultural Economics, Michigan State University, 1986.

Rehber, E., "Vertical Integration in Agriculture and Contract Farming", *Working Paper* No. 46, Food Marketing Policy Center. University of Connecticut, 1998.

Rehber, E., "Vertical Coordination in the Agro-Food Industry and Contract Farming: A Comparative Study of Turkey and the USA", Food Marketing Policy Center, University of Connecticut, *Research Report* No. 52, February 2000.

Roth, R., "Contract Farming Breeds Big Problems for Growers", *Famers'Legal Action Report*, Winter 1992

Runsten, D. and Key, N., "Contract Farming in Developing Countries: Theoretical Aspects and Analysis of Some Mexican Cases", U.N. Economic Commission for Latin America and the Caribbean, Santiago, 1996.

Singh, S., "Contracting out Solutions: Political Economy of Contract Farming in the Indian Punjab", *World Development*, 2002.

Singh, G. and Asokan, S., *Contract Farming in India: Text and Cases*, Oxford & IBH Publishing, New Delhi, 2005

Index